DESIGN AND REALIZATION OF BIPOLAR TRANSISTORS

DESIGN AND MEASUREMENT IN ELECTRONIC ENGINEERING

Series Editors
D. V. Morgan,
Department of Physics, Electronics and Electrical Engineering, University of Wales, Institute of Science and Technology, Cardiff, UK
H. R. Grubin,
Scientific Research Associates Inc., Glastonbury, Connecticut, USA

THYRISTOR DESIGN AND REALIZATION
P. D. Taylor

ELECTRONICS OF MEASURING SYSTEMS
Tran Tien Lang

DESIGN AND REALIZATION OF BIPOLAR TRANSISTORS
Peter Ashburn

DESIGN AND REALIZATION OF BIPOLAR TRANSISTORS

PETER ASHBURN
Southampton University

JOHN WILEY & SONS
Chichester · New York · Brisbane · Toronto · Singapore

Copyright ©1988 by John Wiley & Sons Ltd.

All rights reserved.

No part of this book may be reproduced by any means, or transmitted, or translated into a machine language without the written permission of the publisher

Library of Congress Cataloging-in-Publication Data:
Ashburn, Peter.
  Design and realization of bipolar transistors/Peter Ashburn.
   p. cm.—(Design and measurement in electronic engineering)
  Includes index.
  ISBN 0 471 91700 1
  1. Bipolar transistors—Design and construction. I. Title.
  II. Series.
TK7871.96.B55A84 1988
621.3815'28—dc19   88-2438
                             CIP

British Library Cataloguing in Publication Data:
Ashburn, Peter
  Design and realization of bipolar transistors.
  1. Digital integrated circuits. Bipolar
  transistors. Design.
  I. Title.   II. Series.
   621.381'730422
ISBN 0 471 91700 1

Phototypeset by Dobbie Typesetting Service, Plymouth
Printed and bound in Great Britain by Biddles Ltd, Guildford

To my father, Harold Ashburn

CONTENTS

Series Preface ix
Preface xi
List of Symbols xiii

1 OVERVIEW 1

1.1 Evolution of silicon bipolar technology 1
1.2 Evolution of heterojunction bipolar technology 3
1.3 Operating principles of the bipolar transistor 4

2 BIPOLAR TRANSISTOR THEORY 14

2.1 Introduction 14
2.2 Components of base current 14
2.3 Fundamental equations 16
2.4 Base current 20
2.5 Current gain 24
2.6 Shallow emitters 25
2.7 Heavy doping effects 27
2.8 Extension of the simple theory 38
2.9 Junction breakdown 51

3 BIPOLAR TRANSISTOR MODELS 58

3.1 Transistor modelling 58
3.2 Ebers–Moll model 59
3.3 Small-signal hybrid-Π model 67
3.4 Gummel–Poon model 69
3.5 Modelling the low-current gain 73
3.6 Forward transit time τ_F 74
3.7 Base resistance 80
3.8 Collector/base capacitance 82
3.9 The SPICE bipolar transistor model 83

4 POLYSILICON EMITTERS — 89

- 4.1 Introduction — 89
- 4.2 Basic physics of the polysilicon emitter — 92
- 4.3 Theory of polysilicon emitters — 96
- 4.4 Emitter resistance — 107
- 4.5 Design of practical polysilicon emitter transistors — 109
- 4.6 SIS emitters — 117

5 HETEROJUNCTION EMITTERS — 123

- 5.1 Introduction — 123
- 5.2 Theory of heterojunction emitters — 125
- 5.3 GaAlAs/GaAs heterojunction emitters — 126
- 5.4 Bandgap engineering — 130

6 BIPOLAR INTEGRATED CIRCUIT FABRICATION — 134

- 6.1 Introduction — 134
- 6.2 Buried layer and epitaxy — 135
- 6.3 Isolation — 139
- 6.4 Base — 144
- 6.5 Emitter — 146
- 6.6 Yield problems in bipolar processes — 148
- 6.7 Analogue bipolar processes — 153
- 6.8 Digital bipolar processes — 159
- 6.9 GaAs/GaAlAs heterojunction bipolar processes — 163
- 6.10 BICMOS processes — 166

7 OPTIMIZATION OF HIGH-SPEED BIPOLAR PROCESSES — 174

- 7.1 Introduction — 174
- 7.2 ECL propagation delay expression — 175
- 7.3 Calculation of the electrical parameters — 178
- 7.4 Comparison of conventional and self-aligned processes — 179
- 7.5 Process optimization — 185

APPENDIXES

- 1 Bipolar transistor model parameters — 191
- 2 Fundamental physical constants — 192
- 3 Properties of silicon and gallium arsenide — 192
- 4 Properties of silicon dioxide — 192

INDEX — 193

SERIES PREFACE

The crucial role of design in the engineering industry has been increasingly recognized over recent years, with particular emphasis being placed on this aspect of engineering in first and higher degree work as well as continuing education.

This new series of books concentrates on fundamental aspects of design and measurement in electronic engineering and will involve an international authorship. The authors are sought from scientists and engineers who have made a significant contribution in their field. The books in the series will cover a range of topics at research level and are primarily intended for research and development engineers wishing to gain detailed specialist knowledge of design and measurement in a particular area of electronic engineering. It is assumed that, as a starting point, the reader will have a background degree or equivalent qualification in electrical and electronic engineering, physics or mathematics. In the series no attempt will be made to provide preliminary background material but rather the texts will move directly into the design aspects.

Professor D. V. Morgan
Dr H. R. Grubin

PREFACE

The decade of the 1980s has seen remarkable developments in the design and realization of bipolar transistors. In silicon technology the emergence of polysilicon emitters and self-aligned fabrication techniques has resulted in a considerable improvement in the performance of high-speed digital circuits. Similarly, in gallium arsenide technology GaAs/GaAlAs heterojunctions have developed to the point that self-aligned MSI circuits can be produced with extremely high performance and also reasonable yields. These innovations will not only lead to large improvements in the switching speed of bipolar circuits, but will also open up new applications for bipolar transistors. The integration of GaAs/GaAlAs heterojunction bipolar transistors with opto-electronic devices, such as lasers, is one example of these exciting new possibilities. Heterojunction bipolar technology therefore provides for the first time a viable means of fabricating opto-electronic integrated circuits.

The purpose of this book is to bring together these new developments into a single text which covers both bipolar transistor design and fabrication. The book will also cover the broader topic of the optimization of bipolar devices and processes for high-speed, digital circuits. This will be achieved through the use of a quasi-analytical expression for the gate delay of an ECL logic gate. The book is intended primarily for practising engineers and scientists and for students at the postgraduate level.

In the first chapter the reader is given an overview of silicon and heterojunction technologies and is introduced to the operating principles of the bipolar transistor. A more rigorous and quantitative description of the bipolar transistor is then given in the succeeding two chapters. Chapter 2 deals with the physics of the bipolar transistor and takes the reader through the derivation of an expression for the current gain. Heavy doping effects and recombination via deep levels are covered in detail. Chapter 3 explains the modelling of bipolar transistors and includes detailed descriptions of the Ebers–Moll, Gummel–Poon and SPICE bipolar transistor models. The relationship between the forward transit time τ_F and the cut-off frequency f_T is also explored.

Chapters 4 and 5 explain the operation of important new types of bipolar transistor. Polysilicon emitters are covered in Chapter 4 from both a theoretical and a practical viewpoint. Expressions for the base current and emitter resistance of a polysilicon emitter transistor are derived and compared with results obtained on practical devices. Chapter 5 explains the theory and practice of heterojunction emitters, with particular emphasis on GaAs/GaAlAs heterojunctions.

The last two chapters deal with bipolar transistor fabrication and the optimization of bipolar processes. The key bipolar process building blocks are identified and discussed in detail in Chapter 6. These include buried layer, epitaxy, isolation, base and emitter. Examples are then given of four types of bipolar process: analogue bipolar, high-speed digital, GaAs/GaAlAs heterojunction and BICMOS. The discussion of process optimization in Chapter 7 proceeds through the medium of a quasi-analytical expression for the gate delay of an ECL logic gate in terms of all the time constants of the circuit. The application of the gate delay expression to process optimization is demonstrated by a case study involving the comparison of conventional and self-aligned bipolar processes. Finally, it is used to show how silicon and GaAs/GaAlAs transistors can be designed to give optimum switching speeds in ECL circuits.

Many people have contributed directly and indirectly to the writing of this book, and it would be impossible to find the space to thank them all. Nevertheless, I would like to identify a number of colleagues who have made particularly large contributions to this project. First, acknowledgements should go to Vernon Morgan, the series editor, for suggesting that I should contribute a book to this excellent series. The encouragement of Henri Kemhadjian, the head of the Department of Electronics and Computer Science at Southampton University, should also be acknowledged.

Particular thanks are due to my colleagues in the Microelectronics Group who took the time to give me their comments on the manuscript. These include Greg Parker, Arthur Brunnschweiler, Alan Cuthbertson, Graham Wolstenholme, Ian Post and Nasser Siabi-Shahrivar. A debt of gratitude is also owed to my past and present research students, who have contributed greatly to my understanding of bipolar transistors. These include Bus Soerowirdjo, Alan Cuthbertson, Eng Fong Chor, Graham Wolstenholme, Dylan Evans, David Browne and Ian Post. Thanks are also due to Jane Cross for assistance with the typing of the manuscript.

Finally, no list of acknowledgements would be complete without mention of my wife and family for their support during the extension of this seemingly endless task. I will therefore finish by acknowledging the patience and support of my wife Ann, and children Jennifer and Susan.

Southampton, England Peter Ashburn
November 1987

LIST OF SYMBOLS

A	Area of the emitter/base junction
A_e	Modified Richardson constant
α	Common base current gain
α_F	Forward common base current gain
α_R	Reverse common base current gain
α_T	Base transport factor
b_b	Width of the extrinsic base region of a bipolar transistor
b_c	Width of the buried layer of a bipolar transistor
b_e	Width of the emitter of a bipolar transistor
BV_{CBO}	Bipolar transistor breakdown voltage between the collector and base with the emitter open-circuit
BV_{CEO}	Bipolar transistor breakdown voltage between the emitter and collector with the base open-circuit
β	Common emitter current gain
β_F	Forward common emitter current gain
β_R	Reverse common emitter current gain
C_{DC}	Collector diffusion capacitance
C_{DE}	Emitter diffusion capacitance
C_{JEB}	Emitter/base depletion capacitance
C_{JBC}	Base/collector depletion capacitance
C_{JCI}	Intrinsic collector/base depletion capacitance
C_{JCX}	Extrinsic collector/base depletion capacitance
C_{JCS}	Collector/substrate depletion capacitance
C_μ	Collector/base capacitance in the small-signal hybrid-π model
C_π	Emitter/base capacitance in the small-signal hybrid-π model
C_n	Auger recombination coefficient
C_L	Load capacitance due to interconnections
D_n	Diffusion coefficient of electrons
D_p	Diffusion coefficient of holes
D_{nb}	Diffusion coefficient of electrons in the base
D_{pe}	Diffusion coefficient of holes in the emitter

D_{gb}	Diffusion coefficient of holes in a polysilicon grain boundary
D_{pp}	Diffusion coefficient of holes within a polysilicon grain
d_g	Width of a polysilicon grain
d_{gb}	Width of a polysilicon grain boundary
ΔE_c	Conduction band discontinuity in a heterojunction
ΔE_v	Valence band discontinuity in a heterojunction
ΔE_{gb}	Apparent bandgap narrowing in the base
ΔE_{ge}	Apparent bandgap narrowing in the emitter
ΔV	Logic swing of an ECL gate
δ	Interfacial layer thickness in a polysilicon emitter
E	Electric field
E_{crit}	Critical electric field for avalanche breakdown
E_F	Fermi level
E_{Fn}	Electron quasi-fermi level
E_{Fp}	Hole quasi-fermi level
E_c	Energy level of the conduction band
E_v	Energy level of the valence band
E_g	Semiconductor bandgap
E_i	Intrinsic fermi level
E_t	Energy level of a deep level
e_n	Emission probability for electrons at a deep level
e_p	Emission probability for holes at a deep level
ε_o	Permittivity of free space
ε_r	Relative permittivity or dielectric constant of silicon
F	Fan-out of a logic gate
f_T	Cut-off frequency
f_{TMAX}	Peak value of the cut-off frequency
f_{MAX}	Unity power gain frequency
G_b	Base Gummel number
G_e	Emitter Gummel number
G_n	Electron generation rate
G_p	Hole generation rate
g_m	Transconductance
γ	Emitter efficiency
h	Planck's constant
h_{FE}	Common emitter current gain
h_{FB}	Common base current gain
I_B	Base current
I_C	Collector current
I_E	Emitter current

I_S	Saturation current
I_{pe}	Hole diffusion current in the emitter
I_{ne}	Electron diffusion current at the emitter edge of the base
I_{nc}	Electron diffusion current at the collector edge of the base
I_{rb}	Recombination current in the base
I_{rg}	Recombination current in the emitter/base depletion region
J_n	Electron current density
J_p	Hole current density
J_{NT}	Electron tunnelling current density
J_{PT}	Hole tunnelling current density
K	Boltzmann's constant
χ_e	Effective barrier height for electron tunnelling
χ_h	Effective barrier height for hole tunnelling
χ	Electron affinity
L_{nb}	Electron diffusion length in the base
L_{pe}	Hole diffusion length in the emitter
L_{pp}	Hole diffusion length within a polysilicon grain
l_b	Length of the extrinsic base region of a bipolar transistor
l_c	Length of the buried layer of a bipolar transistor
l_e	Length of the emitter of a bipolar transistor
M	Avalanche breakdown multiplication factor
m	Base current ideality factor
m_e^*	Electron effective mass
m_h^*	Hole effective mass
μ_n	Electron mobility
μ_p	Hole mobility
N_a	Acceptor concentration
N_d	Donor concentration
N_{ab}	Acceptor concentration in the base
N_{dc}	Donor concentration in the collector
N_{de}	Donor concentration in the emitter
N_{deff}	Effective doping concentration, including the effects of bandgap narrowing
N_c	Effective density of states in the conduction band
N_v	Effective density of states in the valence band
N_t	Density of deep levels
N_{it}	Density of traps at the polysilicon/silicon interface
N_{st}	Density of traps at a polysilicon grain boundary
n	Electron concentration
n_b	Electron concentration in the base

n_{bo}	Equilibrium electron concentration in the base
n_i	Intrinsic carrier concentration
n_{io}	Intrinsic carrier concentration in a lightly doped semiconductor
n_{ie}	Intrinsic carrier concentration in a heavily doped emitter
n_{ib}	Intrinsic carrier concentration in a heavily doped base
p	Hole concentration
p_e	Hole concentration in the emitter
p_{eo}	Equilibrium hole concentration in the emitter
Q	Charge
q	Charge on an electron
R_B	Base resistance
R_{BI}	Intrinsic base resistance
R_{BX}	Extrinsic base resistance
R_C	Collector resistance
R_E	Emitter resistance
R_e	Emitter follower resistance in an ECL circuit
R_L	Load resistance in an ECL circuit
R_{SBI}	Sheet resistance of the intrinsic base
R_{SBX}	Sheet resistance of the extrinsic base
R_{SBL}	Sheet resistance of the buried layer
R_{CON}	Contact resistance
S_M	Recombination velocity of a metal contact
S_{PI}	Effective recombination velocity at the polysilicon/silicon interface
S_{EFF}	Effective recombination velocity for a complete polysilicon emitter
S_P	Effective recombination velocity at the edge of the polysilicon layer in a polysilicon emitter
σ_n	Capture cross-section for electrons
σ_p	Capture cross-section for holes
T	Temperature
τ_n	Electron lifetime
τ_p	Hole lifetime
τ_{nb}	Electron lifetime in the base
τ_{pe}	Hole lifetime in the emitter
τ_A	Auger lifetime
τ_F	Forward transit time
τ_R	Reverse transit time
τ_E	Emitter delay
τ_{EBD}	Emitter/base depletion region delay
τ_B	Base transit time
τ_{CBD}	Collector/base depletion region transit time

τ_{RE}	Delay due to the emitter/base and collector/base depletion capacitances
τ_d	Propagation delay
U	Recombination rate
U_n	Electron recombination rate
U_p	Hole recombination rate
V_{BE}	Base/emitter voltage
V_{CB}	Collector/base voltage
V_{CE}	Collector/emitter voltage
V_{AF}	Forward Early voltage
V_{AR}	Reverse Early voltage
V_{bi}	Built-in voltage of a pn junction
v_{th}	Thermal velocity
v_{scl}	Scattering limited velocity
W_B	Basewidth
W_E	Depth of the emitter
W_D	Depletion width
W_{CBD}	Collector/base depletion width

Chapter 1
OVERVIEW

1.1 EVOLUTION OF SILICON BIPOLAR TECHNOLOGY

The bipolar transistor was invented by a team of researchers at the Bell Laboratories, USA, in 1948 [1]. The original transistor was a germanium, point contact device, but in 1949 Shockley published a paper on p–n junctions and junction transistors [2]. These two papers laid the foundations for the modern bipolar transistor, and made possible today's multi-million dollar microelectronics industry.

A large number of innovations and breakthroughs were required to convert the original concept into a practical technology for fabricating VLSI circuits. Among these, diffusion was an important first step, since it allowed thin bases and emitters to be fabricated by diffusing impurities from the vapour phase [3]. The use of epitaxy [4] to produce a thin, single-crystal layer on top of a heavily doped buried layer was also a big step forward. This is illustrated in Figure 1.1, and leads to a substantial reduction in the collector series resistance and the excess stored charge in the collector. Faster switching speeds and improved high-frequency gain were the main consequences of this innovation.

The final stage in the evolution of bipolar technology was the development of the planar process [5], which allowed bipolar transistors and other components, such as resistors, to be fabricated simultaneously. This is clearly necessary if circuits are to be produced on a single silicon chip (i.e. integrated circuits). Figure 1.1 shows the main features of the planar bipolar process.

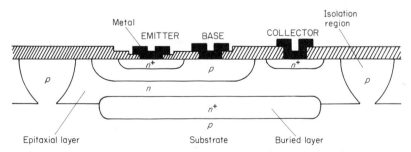

Figure 1.1. Cross-sectional view of an analogue, bipolar, integrated circuit transistor

Electrical isolation between adjacent components is provided by a *p*-type isolation region which is diffused from the surface to intersect the *p*-substrate. For the isolation to be effective, the diffusion must completely surround the device, and the isolation junction must be reverse biased by connecting the substrate to the most negative voltage in the circuit. In the case of resistors, the *n*-type epitaxial layer must also be connected to the most positive voltage in the circuit, which is usually the collector rail V_{cc}. The n^+ diffusion underneath the collector contact is needed to give a low-resistance, ohmic contact.

More recent technology innovations have led to considerable improvements in bipolar transistor and circuit performance. Ion implantation has been used to improve the uniformity and reproducibility of the base [6] and emitter [7] depositions, and also to produce devices with extremely narrow basewidths [8]. Furthermore, the use of polysilicon emitters [9] and self-aligned processing techniques [10] has revolutionized the design and performance of bipolar transistors.

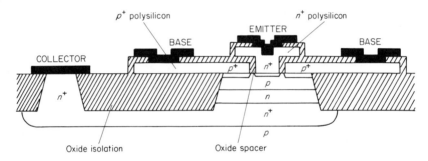

Figure 1.2. Cross-sectional view of a high-speed, bipolar, integrated circuit transistor

Figure 1.2 shows a cross-section of a modern high-speed bipolar transistor. It can be seen that it bears little resemblance to the more traditional transistor in Figure 1.1. Contact to the emitter is made via an n^+ polysilicon layer and to the base via a p^+ polysilicon layer. The emitter and extrinsic base regions are separated by an oxide spacer on the sidewall of the p^+ polysilicon, giving a separation of as little as 0.2 μm. The junction isolation of Figure 1.1 has also been replaced by oxide isolation, which has a much lower parasitic capacitance, and hence is ideal for high-speed applications. Gate delays below 30 ps [11] have been achieved in circuits incorporating bipolar transistors of this type.

The above innovations in technology and device design have been accompanied by similar advances in the design of bipolar logic circuits. The first family of digital integrated circuits was realized using resistor transistor logic (RTL) [12]. This was introduced in 1962, but was quickly followed by diode transistor logic (DTL) [12]. Both these logic families were integrated versions of logic circuits constructed from discrete components.

The main advantage of the planar process was that it allowed the integrated circuit designer new freedom to devise novel device and circuit configurations that had previously not been possible. The first product of this new approach was transistor transistor logic (TTL), which made use of a multi-emitter transistor [13]. Variants of this original TTL logic family are still in wide use today. Integrated injection logic [14] or merged transistor logic [15] was invented in 1972, and made maximum use of the freedom to combine individual components on a single chip. In this logic family, an *npn* and *pnp* transistor were merged together to produce a very simple and very compact logic gate suitable for VLSI applications. At the gate level, the packing density of I^2L is comparable with that of MOS logic families.

For high-speed applications it is important that the bipolar switching transistor is prevented from entering saturation. In this region of operation both emitter and collector junctions are forward biased, with the result that a large amount of charge is injected into the base and collector. This charge must be removed before the transistor can turn off, which leads to very long storage times. In high-speed circuits this is clearly undesirable, and hence logic families such as emitter coupled logic (ECL) have been devised in which the switching transistor is prevented from entering saturation. ECL circuits have the shortest gate delay of any of the commercially available silicon logic families. Circuit operation at 10.4 GHz has been obtained [11] using a self-aligned ECL process.

More recently there has been considerable interest in incorporating bipolar transistors into silicon-gate CMOS processes. The main motivation for moving from CMOS to BICMOS is that bipolar transistors can sink a larger current per unit device area than MOS transistors. They are therefore more effective in driving the large on-chip capacitances that are commonly encountered in digital VLSI systems. In this context, BICMOS logic gates have been designed with a gate delay of 0.71 ns at a load capacitance of 0.85 ps [16]. BICMOS processes also offer the prospect of combining high-speed digital circuits on the same chip as high-performance analogue [17], thereby producing a technology capable of integrating a wide variety of complete electronic systems.

1.2 EVOLUTION OF HETEROJUNCTION BIPOLAR TECHNOLOGY

A heterojunction is a junction formed between two different semiconductors. The original concept of the heterojunction bipolar transistor was proposed by Shockley in 1951 [18]. By using a wide-bandgap semiconductor for the emitter it is possible to increase the gain, decrease the base resistance and reduce the emitter capacitance of the transistor. These improvements in device parameters lead directly to faster switching speeds for digital integrated circuits.

Unfortunately, very little progress was made towards a practical heterojunction technology until the early 1970s. The successful implementation of heterojunction devices depends critically on the ability to join two lattice-matched semiconductors without generating large numbers of defects at the interface. Clearly, this is no simple task, and significant progress was only made after the development of liquid phase epitaxy as a technique for growing GaAs/GaAlAs structures [19]. Since that time two further extremely promising technologies have been developed, namely molecular beam epitaxy (MBE) [20] and metal organic chemical vapour deposition (MOCVD) [21]. These innovations in materials growth techniques have allowed a practical heterojunction technology to develop, based on GaAlAs for the wide-bandgap emitter and GaAs for the smaller bandgap base.

GaAs/GaAlAs heterojunction bipolar technology has been successfully used to fabricate not only bipolar transistors but also MSI high-speed circuits. The technology has evolved in an analogous way to silicon technology, and hence self-aligned processing techniques have been widely used, as illustrated in Figure 1.3. Logic gates with delays of 16.5 ps have been produced using this approach, along with circuits operating at 20 GHz [22]. Another interesting trend in this technology is the emergence of schemes for integrating heterojunction bipolar transistors on the same chip as optical devices such as double-heterostructure lasers [23] and light-emitting diodes [24]. These developments offer the prospect of a viable technology for realizing opto-electronic integrated circuits.

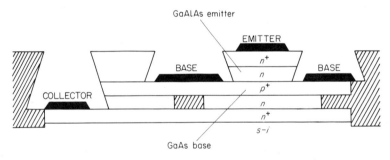

Figure 1.3. Cross-sectional view of a GaAs/GaAlAs heterojunction bipolar transistor

Because of the pre-eminence of silicon technology there is a strong incentive to incorporate heterojunction emitters into existing silicon bipolar processes. Several approaches have been investigated, including the use of semi-insulating polycrystalline silicon (SIPOS) [25] and silicon carbide [26] as the wide-bandgap emitter. This research is at an early stage of development, and considerable work remains to be done before it can be considered as a viable technology.

1.3 OPERATING PRINCIPLES OF THE BIPOLAR TRANSISTOR

For the purposes of understanding the operation of the bipolar transistor the structures in Figures 1.1–1.3 can be considered as essentially one-dimensional, as illustrated in Figure 1.4(a). Although this is clearly an approximation, it is valid over a remarkable range of operating conditions. In practice, it only begins to break down at very high current levels when the series resistances of the

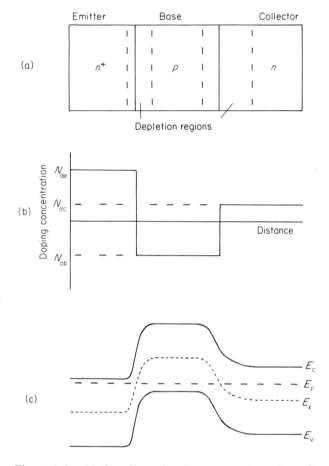

Figure 1.4. (a) One-dimensional representation of an *npn* bipolar transistor; (b) doping profiles for the case of abrupt *pn* junctions; (c) band diagram for a transistor with no applied bias

transistor become important. In the first instance, an idealized bipolar transistor will be assumed in which the doping profiles are uniform, as illustrated in Figure 1.4(b). In practice, this is a good approximation for GaAs/GaAlAs heterojunction transistors, but in silicon transistors the profiles are generally Gaussian. The implications of this deviation from ideality will be considered in the second part of Chapter 2.

The band diagram for our idealized, homojunction bipolar transistor is shown in Figure 1.4(c). In the absence of any applied bias, the Fermi level E_F is constant throughout the device. The Fermi level E_F and the intrinsic Fermi level E_i are related to the carrier concentrations in the emitter, base and collector through the following equations [27]:

$$n = N_c \exp-\left(\frac{E_c - E_F}{KT}\right) = n_i \exp\left(\frac{E_F - E_i}{KT}\right) \quad (1.1)$$

$$P = N_v \exp-\left(\frac{E_F - E_v}{KT}\right) = n_i \exp\left(\frac{E_i - E_F}{KT}\right) \quad (1.2)$$

where

$$E_i = \tfrac{1}{2}(E_c + E_v) + \tfrac{1}{2}KT \ln \frac{N_v}{N_c} \quad (1.3)$$

From equations (1.1) and (1.2) it can be seen that the product of the electron and hole concentrations in a given region of the transistor is a constant:

$$pn = n_i^2 \quad (1.4)$$

The relationship between the doping profiles in Figure 1.4(b) and the band diagram in Figure 1.4(c) is now clear. In particular, the doping concentration in a given region of the device is exponentially related to the separation between the Fermi level and the intrinsic Fermi level $(E_F - E_i)$.

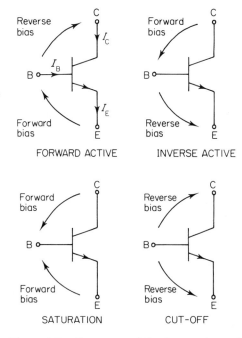

Figure 1.5. Summary of the four regions of operation of a bipolar transistor

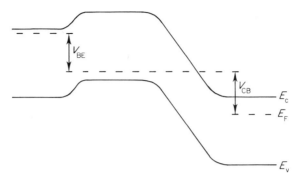

Figure 1.6. Band diagram for a bipolar transistor biased in the forward active region

In order to use the bipolar transistor in practical circuits, external bias must be applied to the emitter/base and collector/base junctions. These two junctions provide four possible bias configurations, as illustrated in Figure 1.5. The forward active mode of operation is the most useful, because in this configuration the gain of the transistor can be exploited to produce current amplification. A forward bias of approximately 0.6 V is applied to the base/emitter junction and a reverse bias to the collector/base junction. The resulting band diagram for this situation is shown in Figure 1.6.

The other bias configurations in Figure 1.5 are also often encountered in practice, particularly in digital circuits. In the inverse (or reverse) active mode, the emitter/base junction is reverse biased and the collector/base forward biased. This arrangement is less useful than the forward active mode because the inverse gain of the transistor is very low, though it is used in I^2L circuits [14,15].

In the cut-off mode both junctions are reverse biased, and hence no current can flow between emitter and collector. The transistor is therefore off, and behaves like an open switch. Conversely, in the saturation mode both junctions are forward biased, which enables a large current to flow between emitter and collector. In this configuration the transistor can be viewed as a closed switch.

The electrical properties of a bipolar transistor can be characterized by a number of electrical parameters, the most important of which is the common emitter current gain β. This is the ratio of collector current to base current, and is given by

$$\beta \equiv h_{FE} = \frac{I_C}{I_B} \qquad (1.5)$$

In a typical commercial transistor the collector current is approximately one hundred times larger than the base current, giving a current gain of one hundred. In order to understand how this important property of the bipolar transistor arises we must consider how it functions when external bias is applied.

In the forward active mode, the forward biasing of the emitter/base junction causes a large number of electrons to be injected from the emitter into the base. A concentration gradient is therefore established in the base, which encourages the electrons to diffuse towards the collector. If the base of the transistor was very wide all the injected electrons would recombine before reaching the collector, and the transistor would merely behave like two back-to-back diodes. However, the essence of the bipolar transistor is that the base is sufficiently narrow that the majority of electrons reach the collector/base junction, where they are swept across into the collector by the large electric field across the reverse biased junction. The small number of electrons that recombine in the base contribute to the base current of the transistor. By making the basewidth comparable with, or smaller than, the diffusion length of electrons, the base current can be made much smaller than the collector current, and hence a sizable current gain achieved. As will be explained in Chapter 2, mechanisms other than recombination in the base also contribute to the base current, and hence this explanation represents an oversimplification. Nevertheless, it is a useful first step in our understanding of the bipolar transistor.

A related electrical parameter to the common emitter current gain is the common base current gain α, which is the ratio of the collector current to the emitter current:

$$\alpha \equiv h_{FB} = \frac{I_C}{I_E} \tag{1.6}$$

From Figure 1.5 the emitter current is given by the sum of the collector and base currents:

$$I_E = I_C + I_B \tag{1.7}$$

It is therefore apparent that α and β are related by:

$$\alpha = \frac{\beta}{1+\beta} \tag{1.8}$$

The common emitter and common base current gains can be measured by biasing the transistor into the forward active region and taking readings of base, emitter and collector current. Three alternative circuit configurations are possible, depending upon which terminal is common between the input and output. These are illustrated in Figure 1.7, and are termed the common emitter, common base and common collector circuit configurations.

The common emitter current gain β is obtained by connecting the transistor in the common emitter configuration illustrated in Figure 1.7(a) and plotting the collector current as a function of collector/emitter voltage, with the base current as a parameter. The resulting characteristics are illustrated in Figure 1.8(a) for a typical practical transistor. At a base current of 10 μA a collector current of 1 mA is obtained, giving a β of 100.

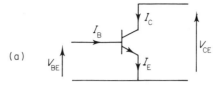

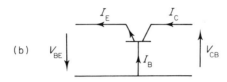

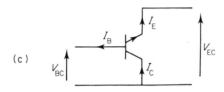

Figure 1.7. The three circuit configurations of a bipolar transistor. (a) Common emitter; (b) common base; (c) Common collector

An alternative measurement method is by means of a Gummel plot, as illustrated in Figure 1.8(b). Here $\ln I_C$ and $\ln I_B$ are plotted as a function of V_{BE}, and β obtained from the ratio of I_C and I_B at a given V_{BE}. It is clear from this figure that both the base and collector current vary exponentially with base/emitter voltage in forward bias, and in fact can be described by:

$$I_C = I_S \exp \frac{qV_{BE}}{KT} \tag{1.9}$$

$$I_B = \frac{I_S}{\beta} \exp \frac{qV_{BE}}{KT} \tag{1.10}$$

The common base current gain α can be measured by connecting the transistor in the common base configuration illustrated in Figure 1.7(b), and plotting the collector current as a function of collector/base voltage, with the emitter current as a parameter. The resulting characteristic is shown in Figure 1.9. At an emitter current of 1 mA a collector current of very nearly 1 mA is obtained, indicating that the common base current gain, is very close to unity.

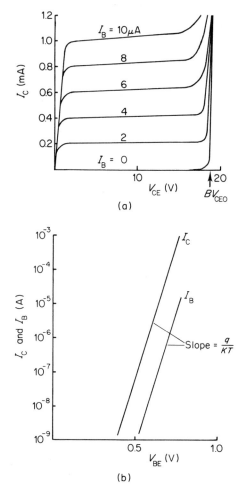

Figure 1.8. Transistor characteristics for a bipolar transistor connected in the common emitter configuration. (a) Output characteristics; (b) Gummel plot

Two further parameters of the bipolar transistor can be obtained from the characteristics in Figures 1.8(a) and 1.9, namely BV_{CEO} and BV_{CBO}. These are the breakdown voltages between collector and emitter with the base open circuit (BV_{CEO}), and between collector and base with the emitter open circuit (BV_{CBO}). In both cases it is the collector/base junction that is breaking down, but, surprisingly, the two breakdown voltages are considerably different. This phenomenon will be explored in more detail in Chapter 2.

The current gain of a bipolar transistor can be used to produce amplification by means of the simple circuit in Figure 1.10. Here the transistor is connected in the common emitter configuration, and DC voltages applied to bias the

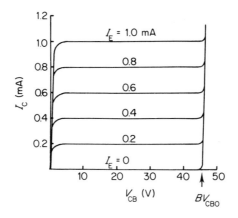

Figure 1.9. Output characteristics of a bipolar transistor connected in the common base configuration

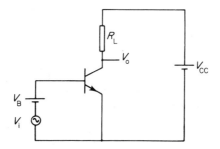

1.10. Simple common emitter bipolar transistor amplifier

transistor into the forward active region of operation. With this arrangement, the DC collector current is h_{FE} times larger than the DC base current. Amplification is obtained by superimposing a small AC signal onto the base current and measuring the AC output signal at the collector of the transistor. The AC component of the collector current is h_{fe} times larger than the AC component of the base current, where h_{fe} is the small signal current gain. If the frequency of the AC signal is low enough that the capacitance of the bipolar transistor is unimportant, then h_{fe} is approximately equal to h_{FE}, and considerable amplification is achieved.

Another important electrical parameter of a bipolar transistor is its transconductance, defined by

$$g_m = \frac{dI_C}{dV_{BE}} \qquad (1.11)$$

The relationship between collector current and base/emitter voltage in a bipolar transistor is exponential, as illustrated in equation (1.9). It is clear therefore that a large change in collector current will result from a very small change in base/emitter voltage. For example, a change of 0.1 V in the base/emitter voltage leads to an increase in collector current by a factor of nearly 50. This large transconductance is very valuable, and it enables logic gates to be designed with logic swings as small as 0.225 V [28]. This in turn makes possible very fast switching speeds when compared with MOS logic gates.

References

1. J. Bardeen and W. H. Brattain, 'The transistor, a semiconductor triode', *Phys. Rev.*, **74**, 230 (1948).
2. W. Shockley, 'The theory of $p-n$ junctions in semiconductors and $p-n$ junction transistors', *Bell Syst. Tech. Jnl*, **28**, 435 (1949).
3. M. Tanenbaum and D. E. Thomas, 'Diffused emitter and base silicon transistor', *Bell Syst. Tech. Jnl*, **35**, 1 (1956).
4. H. C. Theuerer, J. J. Kleimack, H. H. Loar and H. Christenson, 'Epitaxial diffused transistors', *Proc. IRE*, **48**, 1642 (1960).
5. J. A. Hoerni, 'Planar silicon transistor and diodes', IRE Electron Devices Meeting, Washington DC (1960).
6. P. Ashburn, C. J. Bull, K. H. Nicholas and G. R. Booker, 'Effects of dislocations in silicon transistors with implanted bases, *Solid State Electronics*, **20**, 731 (1977).
7. C. Bull, P. Ashburn, G. R. Booker and K. H. Nicholas, 'Effects of dislocations in silicon transistors with implanted emitters', *Solid State Electronics*, **22**, 95 (1979).
8. J. Graul, H. Kaiser, W. Wilhelm and H. Ryssel, 'Bipolar high-speed, low-power gates with double implanted transistors', *IEEE Jnl Solid State Circuits*, **SC10**, 201 (1975).
9. J. Graul, A. Glasl and H. Murrmann, 'High performance transistors with arsenic-implanted polysil emitters', *IEEE Jnl Solid State Circuits*, **SC11**, 491 (1976).
10. T. H. Ning, R. D. Isaac, P. M. Solomon, D. D. Tang, H. Yu, G. C. Feth and S. K. Wiedmann, 'Self-aligned bipolar transistors for high-performance and low power delay VLSI', *IEEE Trans. Electron. Devices*, **ED28**, 1010 (1981)
11. T. Sakai, S. Konaka, Y. Yamamoto and M. Suzuki, 'Prospects of SST technology for high-speed LSI', *IEDM Technical Digest*, 18 (1985).
12. D. J. Hamilton and W. G. Howard, *Basic Integrated Circuit Engineering*, McGraw-Hill, New York (1975).
13. D. A. Hodges and H. G. Jackson, *Analysis and Design of Digital Integrated Circuits*, McGraw-Hill, New York (1983).
14. K. Hart and A. Slob, 'Integrated injection logic: a new approach to LSI', *IEEE Jnl Solid State Circuits*, **SC7**, 346 (1972).
15. H. H. Berger and S. K. Wiedmann, 'Merged transistor logic: a low cost bipolar logic concept', *IEEE Jnl Solid State Circuits*, **SC7**, 340 (1972).
16. H. Higuchi, G. Kitsukawa, T. Ikeda, Y. Nishio, N. Sasaki and J. Ogiue, 'Performance and structures of scaled-down bipolar devices merged with CMOSFETS', *IEDM Technical Digest*, 694 (1984).
17. S. Krishna, J. Kuo and I. S. Gaeta, 'An analog technology integrates bipolar, CMOS, and high voltage DMOS transistors', *IEEE Trans. Electron. Devices*, **ED31**, 89 (1984).
18. W. Shockley, US patent 2,569,347 (1951).
19. W. P. Dumke, J. M. Woodall and V. L. Rideout, 'GaAs–GaAlAs heterojunction transistors for high frequency operation', *Solid State Electronics*, **15**, 1339 (1972).

20. A. Y. Cho and J. R. Arthur, 'Molecular beam epitaxy', *Prog. Solid State Chem.*, **10**, Pt 3, 157 (1975).
21. R. D. Dupuis, L. A. Moudy and P. D. Dapkus, 'Preparation and properties of $Ga_{1-x}Al_xAs$–GaAs heterojunctions grown by metal organic chemical vapour deposition', *Gallium Arsenide and Related Compounds 1978, Institute of Physics Conference Series*, **45**, 1 (1979).
22. K. W. Wang, P. M. Asbeck, M. F. Chang, G. J. Sullivan and D. L. Miller, 'High speed circuits for lightwave communication systems implemented with AlGaAs/GaAs heterojunction bipolar transistors', *Bipolar Circuits and Technology Meeting Digest*, 142 (1987).
23. J. Katz, N. Bar-Chaim, P. C. Chen, S. Margalit, I. Urij, D. Wilt, M. Yust and A. Yariv, 'A monolithic integration of a GaAs/GaAlAs bipolar transistor and a heterostructure laser', *App. Phys. Lett.*, **37**, 211 (1980).
24. H. Beneking, N. Grote and M. N. Svilans, 'Monolithic GaAlAs/GaAs infra-red to visible wavelength converter with optical power amplification', *IEEE Trans. Electron. Devices*, **ED28**, 404 (1981).
25. T. Matsushita, N. Oh-uchi, H. Hahashi and H. Yamoto, 'A silicon heterojunction transistor', *App. Phys. Lett.*, **35**, 549 (1979).
26. K. Sasaki, M. M. Rahman and S. Furukawa, 'An amorphous SiC:H emitter heterojunction bipolar transistor', *IEEE Electron. Device Lett.*, **EDL6**, 311 (1985).
27. S. M. Sze, *Physics of Semiconductor Devices*, John Wiley, New York (1981).
28. M. Suzuki, H. Hagimoto, H. Ichino and S. Konaka, 'A 9GHz frequency divider using Si bipolar super self-aligned process technology', *IEEE Electron, Device Lett.*, **EDL6**, 181 (1985).

Chapter 2
BIPOLAR TRANSISTOR THEORY

2.1 INTRODUCTION

In this chapter we will develop a quantitative theory for the DC characteristics of a bipolar transistor. The approach taken will be to initially derive an approximate analytical expression for the common emitter current gain, using a simplified description of the bipolar transistor. This will allow the physical principles of the device operation to be clearly explained without resorting to undue mathematical complexity.

In the second part of the chapter the deficiencies of the simple theory will be outlined, and a more rigorous description of the transistor behaviour produced. This requires the incorporation of additional physical mechanisms such as heavy doping effects, recombination in the emitter/base depletion region, high-level injection, basewidth modulation and junction breakdown. Many of these mechanisms are difficult to model analytically, and hence as the chapter progresses increasing use will be made of computer-generated data. This approach is entirely appropriate for the modern process and device engineer, since computer-aided numerical simulation is fast becoming an essential part of device design.

2.2 COMPONENTS OF BASE CURRENT

In Chapter 1 we reasoned that the base current of a bipolar transistor was determined by recombination of injected electrons in the base. Although this reasoning is correct, it is by no means the only source of base current. In this section the other possible components of base current will therefore be described.

Figure 2.1 shows a schematic illustration of a bipolar transistor operating in the forward active region, that is, with the emitter/base junction forward biased and the collector/base reverse biased. The forward biasing of the emitter/base junction causes electrons to be injected into the base and likewise holes into the emitter. Considering the electron current first, as electrons leave the emitter some inevitably recombine with holes in the emitter/base depletion layer. This gives rise to a recombination current I_{rg}. The remaining electrons reach the edge of the emitter/base depletion region where they become minority

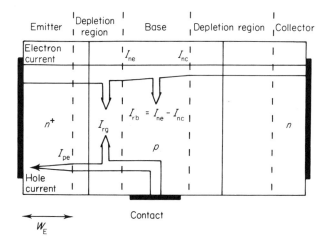

Figure 2.1. Current components in an *npn* bipolar transistor operating in the forward active mode

carriers. A concentration gradient of electrons is established in the base, which encourages them to diffuse towards the collector. The electron diffusion current at the left-hand edge of the neutral base region is defined as I_{ne}. Further electrons recombine with holes in the base, so that the electron diffusion current at the right-hand edge of the base I_{nc} is smaller than I_{ne}. The difference between these two currents is the recombination current in the base I_{rb}. Negligible recombination occurs in the collector/base depletion region because of the high electric field across this reverse-biased junction. Similarly, once the electrons reach the *n*-type collector they become majority carriers, and hence no further recombination occurs.

The hole current injected from the base into the emitter is also shown in Figure 2.1. As with the electron current, a small fraction of the injected holes recombine in the emitter/base depletion region, giving rise to the recombination current I_{rg}. The remaining holes progress to the emitter where they become minority carriers, and are able to diffuse towards the emitter contact. The hole diffusion current at the edge of the emitter/base depletion region is defined as I_{pe}. At this point two situations can arise, depending upon the thickness of the emitter W_E with respect to the hole diffusion length. If the emitter is very thick, all injected holes recombine with electrons before reaching the metal contact. In this case, the metal contact has no effect on the hole diffusion current and hence on the gain. This is the situation illustrated in Figure 2.1. Alternatively, if the emitter is very thin, the majority of the holes reach the contact without recombining. In this case, recombination occurs at the contact, and the properties of the contact have a strong influence on I_{pe}. The former situation will be considered in the subsequent sections and the latter in Section 2.6.

By inspection of Figure 2.1 we can write the components of the emitter, collector and base currents:

$$I_E = I_{ne} + I_{rg} + I_{pe} \tag{2.1}$$

$$I_C = I_{nc} \tag{2.2}$$

$$I_B = I_E - I_C = I_{pe} + I_{rg} + I_{rb} \tag{2.3}$$

Strictly speaking, an additional current component can arise from the leakage current of the reverse-biased collector/base junction. However, in practical devices this current is of the order of 1 nA/cm^2 and hence can be neglected.

At this point we are in a position to define two additional parameters of the bipolar transistor. The emitter efficiency γ is defined as the ratio of the electron current injected into the base to the total emitter current:

$$\gamma = \frac{I_{ne}}{I_{ne} + I_{rg} + I_{pe}} \tag{2.4}$$

From this equation we can see that an efficient emitter is one in which I_{rg} and I_{pe} are much smaller than I_{ne}. Intuitively, we would expect I_{pe} to be smaller than I_{ne} only if the number of holes in the device was smaller than the number of electrons. This reasoning is correct, and leads to the design criterion that the emitter doping must be much larger than the base doping in order to produce an efficient emitter.

The efficiency of the base is defined by the transport factor α_T, which is the ratio of the electron current reaching the collector to that injected from the emitter:

$$\alpha_T = \frac{I_{nc}}{I_{ne}} \tag{2.5}$$

An efficient base is obtained when I_{nc} is nearly equal to I_{ne}, a situation which arises when the base is very narrow.

Finally, from equations (2.1)–(2.5) it can be seen that the common base current gain is given by:

$$\alpha = \gamma \alpha_T \tag{2.6}$$

2.3 FUNDAMENTAL EQUATIONS

The fundamental equations for describing the transport of carriers in a semiconductor under non-equilibrium conditions are the electron and hole continuity equations [1]:

$$\frac{\partial n}{\partial t} = G_n - U_n + \frac{1}{q} \nabla \cdot \mathbf{J}_n \tag{2.7}$$

$$\frac{\partial p}{\partial t} = G_p - U_p - \frac{1}{q} \nabla \cdot \mathbf{J}_p \tag{2.8}$$

where $\mathbf{J}_n$ and $\mathbf{J}_p$ are the electron and hole current densities, G_n and G_p the electron and hole generation rates ($m^{-3}s^{-1}$) due to external excitation and U_n and U_p the electron and hole recombination rates.

The solutions of these equations under appropriate boundary conditions give the electron and hole concentrations as a function of space and time. In order to arrive at an explicit solution, expressions for the current densities $\mathbf{J}_n$ and $\mathbf{J}_p$ in terms of the electron and hole concentrations are needed. These equations can readily be derived by expressing the current as the sum of a diffusion and drift term:

$$\mathbf{J}_n = qD_n \nabla n + qn\mu_n E \qquad (2.9)$$

$$\mathbf{J}_p = -qD_p \nabla p + qp\mu_p E \qquad (2.10)$$

Here the diffusion current is proportional to the gradient of the carrier concentration, indicating that carriers flow from a region of high concentration to one of low concentration. The constants D_n and D_p are the diffusion coefficients or diffusivities, and are related to the mobilities μ_n and μ_p through the Einstein relations:

$$D_n = \mu_n \frac{KT}{q} \qquad (2.11)$$

$$D_p = \mu_p \frac{KT}{q} \qquad (2.12)$$

In general, a further equation is needed in order to specify the electric field $\mathbf{E}$. Poisson's equation provides this expression, and relates the electric field to the charge density per unit volume ρ:

$$\nabla \cdot \mathbf{E} = \frac{\rho}{\varepsilon_0 \varepsilon_r} \qquad (2.13)$$

where ε_0 is the permittivity of free space and ε_r the relative permittivity or dielectric constant. For specific problems, the charge density ρ can be expressed in terms of the electron and hole concentration, thereby providing a complete set of equations for solution.

2.3.1 Assumptions

The above equations allow a complete three-dimensional solution to be obtained for the gain of a bipolar transistor. Fortunately, however, such a rigorous analysis is not necessary, since the electrical characteristics of most practical bipolar transistors can be reasonably accurately described by a one-dimensional solution. Furthermore, a considerable simplification of the mathematics can be obtained if a number of assumptions are made:

(1) Steady-state conditions prevail: i.e.

$$\frac{\partial n}{\partial t} = \frac{\partial p}{\partial t} = 0 \qquad (2.14)$$

(2) There is no external generation of carriers: i.e.

$$G_n = G_p = 0 \qquad (2.15)$$

(3) All regions of the device are uniformly doped, as shown in Figure 2.2(b). This implies that there is no built-in electric field.

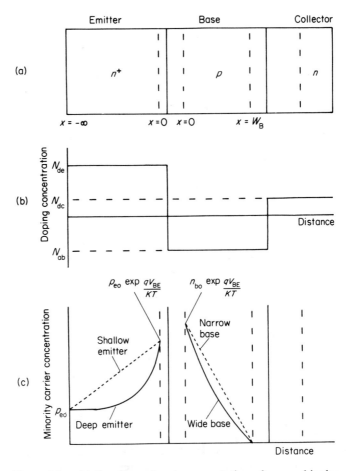

Figure 2.2. (a) One-dimensional representation of an *npn* bipolar transistor; (b) doping profiles for the case of abrupt *pn* junctions; (c) schematic representation of the minority carrier distributions in the emitter and base of an *npn* bipolar transistor operating in the forward active region

(4) The conductivities in the bulk semiconductor regions are high enough to ensure that all the applied voltage is dropped across the depletion regions. This assumption, when taken together with assumption (3), indicates that carriers in the bulk regions of the device move under the influence of diffusion only. This provides a considerable simplification of the mathematics, since the electric field in equations (2.9) and (2.10) can be set to zero, thereby eliminating the requirement for a solution of Poisson's equation.

(5) No generation or recombination of carriers occurs in the depletion regions of the device. This assumption is required in order that simple boundary conditions can be established for the continuity equations. It is a reasonable approximation in most circumstances, but problems can arise in some types of device, as will be discussed later in this chapter.

(6) Low-level injection conditions prevail. That is, the number of electrons injected from the emitter into the base is small compared with the doping concentration in the base. This assumption is valid at low collector currents, but is violated at high currents, as will be considered later in this chapter.

For the bipolar transistor in Figure 2.2, application of the above approximations yields simplified expressions for the electron and hole diffusion current densities:

$$J_n = qD_{nb}\frac{dn_b}{dx} \qquad (2.16)$$

$$J_p = -qD_{pe}\frac{dp_e}{dx} \qquad (2.17)$$

Here the subscripts b and e refer to base and emitter, respectively. A similar procedure can also be used to simplify the continuity equations:

$$D_{nb}\frac{d^2 n_b}{dx^2} - \frac{(n_b - n_{bo})}{\tau_{nb}} = 0 \qquad (2.18)$$

$$D_{pe}\frac{d^2 p_e}{dx^2} - \frac{(p_e - p_{eo})}{\tau_{pe}} = 0 \qquad (2.19)$$

In equations (2.18) and (2.19) the recombination rates U_n and U_p have been represented by:

$$U_n = \frac{n_b - n_{bo}}{\tau_{nb}} \qquad (2.20)$$

$$U_p = \frac{p_e - p_{eo}}{\tau_{pe}} \qquad (2.21)$$

where τ_{nb} and τ_{pe} are the minority carrier lifetimes in the base and emitter, and n_{bo} and p_{eo} the thermal equilibrium values of the minority carrier concentrations in the base and emitter. The terms $(n_b - n_{bo})$ and $(p_e - p_{eo})$ therefore represent the excess minority carrier concentrations.

2.4 BASE CURRENT

2.4.1 Hole Diffusion Current

The most important component of the base current in the majority of bipolar transistors is the hole diffusion current I_{pe}. This can be calculated by solving equations (2.17) and (2.19) under appropriate boundary conditions. For the time being, we will assume that all the injected holes recombine before reaching the emitter contact, or, in other words, that the emitter depth W_E is much greater than the diffusion length of holes in the emitter. This boundary condition can be expressed mathematically as:

$$p_e(-\infty) = p_{eo} \qquad (2.22)$$

The second boundary condition defines the hole concentration at the emitter side of the emitter/base depletion region. In order to derive an equation for this quantity we need to know how the electron and hole concentrations vary across the depletion region. This information is obtained by invoking the simplifying assumption of quasi-equilibrium [2], which states that the pn product is constant throughout the forward-biased depletion region. An alternative way of starting this assumption is through electron and hole quasi-Fermi levels E_{Fn} and E_{Fp}. These define the carrier concentrations under non-equilibrium conditions, and are given by the following equations:

$$n = n_i \exp \frac{(E_{Fn} - E_i)}{KT} \qquad (2.23)$$

$$p = n_i \exp \frac{(E_i - E_{Fp})}{KT} \qquad (2.24)$$

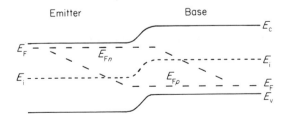

Figure 2.3. Band diagram showing the quasi-Fermi levels for an emitter/base junction under forward bias

The quasi-equilibrium assumption is equivalent to assuming that the quasi-Fermi levels are constant throughout the depletion region, as illustrated in Figure 2.3. From this figure it is clear that the quasi-Fermi levels are related to the base/emitter voltage through:

$$qV_{BE} = E_{Fn} - E_{Fp} \tag{2.25}$$

Combining equations (2.24) and (2.25) with the equation for the equilibrium hole concentration yields the required boundary condition:

$$p_e(0) = p_{eo} \exp \frac{qV_{BE}}{KT} \tag{2.26}$$

This equation indicates that the hole concentration at the emitter edge of the emitter/base depletion region is exponentially related to the base/emitter voltage.

The solution of equation (2.19) with the boundary conditions in equations (2.22) and (2.26) is:

$$p_e - p_{eo} = p_{eo} \left(\exp \frac{qV_{BE}}{KT} - 1 \right) \exp \frac{x}{L_{pe}} \tag{2.27}$$

where

$$L_{pe} = (D_{pe} \tau_{pe})^{1/2} \tag{2.28}$$

L_{pe} is the hole diffusion length in the emitter. Equation (2.27) indicates that the excess hole concentration in the emitter decays exponentially with distance towards the emitter contact, as illustrated by the solid line in Figure 2.2(c). The hole diffusion current I_{pe} can be calculated from equations (2.17) and (2.27), and is given by:

$$I_{pe} = -qAD_{pe} \left(\frac{dp_e}{dx} \right)_{x=0} = -\frac{qAD_{pe}n_i^2}{L_{pe}N_{de}} \exp \frac{qV_{BE}}{KT} \tag{2.29}$$

$$p_{eo} n_{eo} = p_{eo} N_{de} = n_i^2 \tag{2.30}$$

where it has been assumed that $V_{BE} \gg KT/q$ and N_{de} is the donor concentration in the emitter.

2.4.2 Recombination Current in the Base

The recombination current in the base I_{rb} can be calculated by solving equations (2.16) and (2.18) under appropriate boundary conditions. At the base edge of the emitter/base depletion region the electron concentration is given by an equation analogous to equation (2.26):

$$n_b(0) = n_{bo} \exp \frac{qV_{BE}}{KT} \qquad (2.31)$$

The second boundary condition defines the electron concentration at the base edge of the collector/base depletion region, and is given by:

$$n_b(W_B) = n_{bo} \exp -\frac{qV_{CB}}{KT} \simeq 0 \qquad (2.32)$$

This equation indicates that all minority carrier electrons in the vicinity of the reverse-biased collector/base junction are swept into the collector by the high electric field.

The solution of equation (2.18) with the boundary conditions in equations (2.31) and (2.32) is:

$$n_b = \frac{n_{bo}[\exp(qV_{BE}/KT)-1]\sinh\left(\frac{W_B-x}{L_{nb}}\right)}{\sinh(W_B/L_{nb})} + n_{bo}\left[1 - \frac{\sinh(x/L_{nb})}{\sinh(W_B/L_{nb})}\right] \qquad (2.33)$$

where

$$L_{nb} = (D_{nb}\tau_{nb})^{1/2} \qquad (2.34)$$

L_{nb} is the electron diffusion length in the base. For practical values of forward bias, such that $V_{BE} \gg KT/q$, the second term in equation (2.33) can be neglected, giving:

$$n_b = \frac{n_{bo}[\exp(qV_{BE}/KT)]\sinh\left(\frac{W_B-x}{L_{nb}}\right)}{\sinh(W_B/L_{nb})} \qquad (2.35)$$

The electron distribution in the base corresponding to equation (2.35) is illustrated schematically by the solid line in Figure 2.2(c).

In high-speed bipolar transistors the basewidth W_B is made as small as possible (typically 0.1 μm) in order to minimize the base transit time. This compares with a typical electron diffusion length in the base of 10 μm. It is clear that in this case $W_B \ll L_{nb}$, and equation (2.35) can be simplified to:

$$n_b = n_{bo}\left(1 - \frac{x}{W_B}\right)\exp\frac{qV_{BE}}{KT} \qquad (2.36)$$

This is a linear distribution, and is illustrated by the dashed line in Figure 2.2(c). Equation (2.36) represents the case of negligible recombination in the base, and is applicable to the majority of integrated circuit bipolar transistors.

In power bipolar transistors, recombination in the base is in general not negligible, and the recombination current in the base I_{rb} is given by:

$$I_{rb} = I_{ne} - I_{nc} \qquad (2.37)$$

The electron diffusion current at the edge of the emitter/base depletion region I_{ne} can be calculated from equations (2.16) and (2.35):

$$I_{ne} = qAD_{nb}\left(\frac{dn_b}{dx}\right)_{x=0} = -\frac{qAD_{nb}n_{bo}}{L_{nb}}\coth\frac{W_B}{L_{nb}}\exp\frac{qV_{BE}}{KT} \qquad (2.38)$$

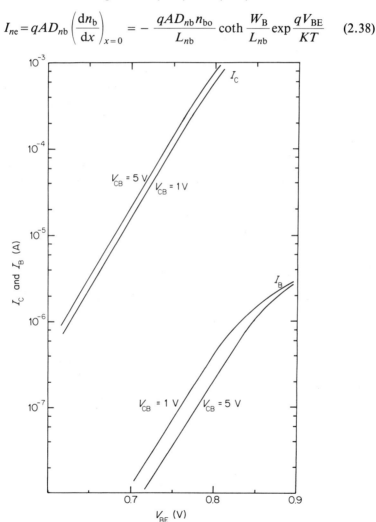

Figure 2.4. Gummel plots for a practical silicon bipolar transistor showing the effects of the application of a collector/base reverse bias

Similarly, the electron diffusion current at the edge of the collector/base depletion region I_{nc} can be calculated:

$$I_{nc} = qAD_{nb}\left(\frac{dn_b}{dx}\right)_{x=W_B} = -\frac{qAD_{nb}n_{bo}}{L_{nb}\sinh(W_B/L_{nb})}\exp\frac{qV_{BE}}{KT} \quad (2.39)$$

The recombination current in the base is therefore given by:

$$I_{rb} = -\frac{qAD_{nb}n_{bo}}{L_{nb}\sinh(W_B/L_{nb})}\left(\cosh\frac{W_B}{L_{nb}} - 1\right)\exp\frac{qV_{BE}}{KT} \quad (2.40)$$

As expected, this equation indicates that the recombination current in the base is a function of the basewidth W_B of the transistor.

In practical bipolar transistors the importance of recombination in the base can be assessed by varying the basewidth of the transistor and looking for changes in base current. This can most easily be achieved by varying the collector/base voltage, thereby modulating the depletion width. Although the majority of the depletion region extends into the collector, some penetration into the base can generally be obtained, particularly if a buried layer is present. The results of this procedure are illustrated in Figure 2.4 for a practical transistor in which recombination in the base is significant. Increasing the collector/base voltage from 1 V to 5 V causes a decrease in base current by a factor of approximately two, thereby indicating the importance of recombination in the base in this particular transistor.

2.5 CURRENT GAIN

The common emitter current gain of a bipolar transistor is given by the ratio of the collector current to the base current. From equations (2.2) and (2.39) the collector current is given by:

$$I_c = -\frac{qAD_{nb}n_i^2}{N_{ab}L_{nb}\sinh/(W_B/L_{nb})}\exp\frac{qV_{BE}}{KT} \quad (2.41)$$

where

$$n_{bo}p_{bo} = n_{bo}N_{ab} = n_i^2 \quad (2.42)$$

As discussed in the previous section, recombination in the base is negligible in the great majority of integrated circuit transistors. In the remainder of this chapter, recombination in the base will therefore be neglected. Under these circumstances, $W_B \ll L_{nb}$, and equation (2.41) simplifies to:

$$I_C = -\frac{qAD_{nb}n_i^2}{W_B N_{ab}}\exp\frac{qV_{BE}}{KT} \quad (2.43)$$

In the absence of recombination in the base and in the emitter/base depletion region the base current is given by equations (2.3) and (2.29):

$$I_\text{B} = - \frac{qAD_{pe}n_\text{i}^2}{L_{pe}N_\text{de}} \exp\frac{qV_\text{BE}}{KT} \qquad (2.44)$$

The common emitter current gain is obtained by combining equations (2.43) and (2.44):

$$\beta = \frac{D_\text{nb}L_{pe}N_\text{de}}{D_{pe}W_\text{B}N_\text{ab}} \qquad (2.45)$$

This equation, although a considerable simplification of the situation obtained in practice, illustrates some important design principles of the bipolar transistor. In particular, it is immediately apparent that the gain depends strongly on the ratio of the doping concentrations in the emitter and base N_de/N_ab. In order to obtain a high gain the doping concentration in the emitter should be as high as possible. Similar reasoning also suggests that the doping concentration in the base should be as low as possible. However, in practice, it is necessary to take other important electrical parameters into account. As will be explained in later chapters, the base resistance is critical in determining the switching speed of bipolar circuits, and hence it is desirable to maintain its value as low as possible. This clearly conflicts with the requirement for a high gain, and in practice an engineering compromise is arrived at, in which a gain of approximately 100 is chosen.

Equation (2.45) shows that the gain depends on the basewidth W_B as well as the doping concentration in the base. In practice, it is the product of these two terms that controls the current gain, and this is often referred to as the base Gummel number G_b of the transistor:

$$G_\text{b} = \frac{W_\text{B}N_\text{ab}}{D_{nb}} \qquad (2.46)$$

2.6 SHALLOW EMITTERS

In many high-speed bipolar transistors the emitter/base junction is made extremely shallow in order to minimize peripheral emitter/base capacitance. In this type of device the junction depth is likely to be comparable with, or smaller than, the diffusion length for holes in the emitter. The majority of injected holes therefore diffuse across the emitter without recombining and congregate at the contact. In these circumstances the physical properties of the emitter contact have a strong influence on the current gain.

The base current of a shallow emitter transistor can be calculated by replacing the boundary condition used in Section 2.4 (equation (2.22)) with a surface boundary condition, which defines the recombination rate at the surface [2]:

$$J_p(W_E) = qD_{pe}\left(\frac{dp_e}{dx}\right)_{x=W_E} = qS_M(p_e(W_E) - p_{eo}) \quad (2.47)$$

This equation states that all minority carriers which reach the surface must recombine there. As with bulk recombination, the recombination rate at the surface is proportional to the excess minority carrier concentration ($p_e - p_{eo}$). The proportionality constant S_M is referred to as the surface recombination velocity, and has units of cm s^{-1}. For convenience and ease of comparison with equations in the literature, the metal contact has been redefined to occur at $x = W_E$, as shown in Figure 2.5.

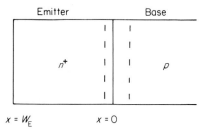

Figure 2.5. Definition of parameters for a shallow emitter

Equation (2.47) is a general boundary condition, and can be used to describe any semiconductor surface. For the special case of an ohmic contact, however, the surface recombination velocity is extremely high, and in most cases can be assumed to be infinite. From equation (2.47) it can be seen that this is equivalent to assuming that the excess minority carrier concentration at the surface is zero.

The base current of a shallow emitter transistor can be calculated by solving equation (2.19) for the hole distribution, using equation (2.47) as a boundary condition:

$$p_e - p_{eo} = p_{eo} \exp\frac{qV_{BE}}{KT}\left[1 - \frac{x}{W_E}\right] \quad (2.48)$$

In deriving this equation it has been assumed that the emitter depth is much less than the hole diffusion length ($W_E \ll L_{pe}$) and also that $S_M \gg D_{pe}/L_{pe}$. This latter approximation is equivalent to assuming that the emitter contact is ohmic, as discussed above. It is clear from equation (2.48) that the excess hole concentration varies linearly with distance, as illustrated by the dashed line in Figure 2.2(c).

In the absence of recombination in the base and in the emitter/base depletion region the base current can be calculated as:

$$I_B = I_{pe} = \frac{qAD_{pe}n_i^2}{W_E N_{de}} \exp \frac{qV_{BE}}{KT} \tag{2.49}$$

Similarly, the common emitter current gain is given by:

$$\beta = \frac{D_{nb} W_E N_{de}}{D_{pe} W_B N_{ab}} \tag{2.50}$$

This equation indicates that the common emitter current gain decreases as the emitter/base junction depth W_E decreases. This degradation of the current gain imposes a practical limit to the extent that the emitter/base junction depth can be reduced. It will be shown in later chapters that shallow emitter/base junctions are desirable in small geometry bipolar transistors in order to minimize the peripheral emitter/base capacitance. This gain degradation is therefore a serious problem in the scaling of high-speed bipolar transistors. Polysilicon emitters and heterojunction emitters, which will be discussed in Chapters 4 and 5, provide two alternative solutions to this problem.

2.7 HEAVY DOPING EFFECTS

The simple analysis in the previous sections clearly indicates the desirability of using a very high doping concentration in the emitter of a bipolar transistor. Unfortunately, in reality the promised advantages of a highly doped emitter do not materialize, particularly in silicon transistors. For example, the gain is significantly smaller than predicted by equation (2.45) [3], and is also strongly temperature dependent [4]. These discrepancies between theory and experiment can be accounted for by heavy doping effects which have not been taken into account in the simple theory. For convenience, they can be characterized by three separate but related mechanisms, namely mobility degradation at high doping concentrations [5], bandgap narrowing [6] and Auger recombination [7].

2.7.1 Mobility

Mobility is a measure of the time interval between collisions for a carrier moving through a semiconductor lattice. The two most important collision mechanisms are lattice and impurity scattering, and the total mobility is given by the sum of the probabilities of collisions due to these individual mechanisms:

$$\frac{1}{\mu} = \frac{1}{\mu_I} + \frac{1}{\mu_L} \tag{2.51}$$

Lattice scattering is caused by collisions between carriers and the atoms of the semiconductor lattice. These lattice atoms are displaced from their lattice sites by thermal vibration, which has the effect of disrupting the perfect periodicity of the semiconductor lattice. Since thermal motion increases with temperature it is not surprising to discover that μ_L decreases with temperature. In fact it can be shown [8] that μ_L varies as $T^{-3/2}$.

Impurity scattering is caused by collisions between carriers and impurity atoms in the semiconductor lattice. As will be discussed in the following section, impurity or dopant atoms have the effect of disrupting the perfect periodicity of the semiconductor lattice, and the amount of disruption increases with impurity concentration. The mobility due to impurity scattering μ_I therefore decreases with increasing impurity concentration.

Experimental values of mobility as a function of impurity concentration [1,9,10] are shown in Figure 2.6 for both silicon and gallium arsenide. It can be seen that the mobility is highest at low impurity concentrations where lattice scattering is the dominant mechanism. At higher impurity concentrations both electron and hole mobilities continuously decrease with increasing dopant concentration. For silicon at impurity concentrations above 10^{19} cm^{-3} the

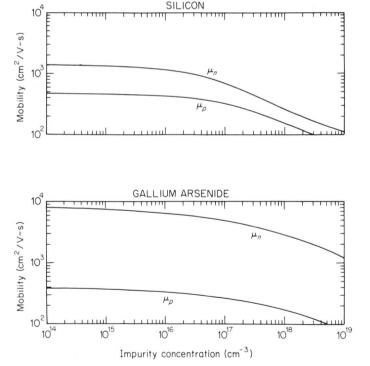

Figure 2.6. Measured values of majority carrier mobility as a function of impurity concentration for silicon and gallium arsenide (after Sze [1]. Reproduced with permission)

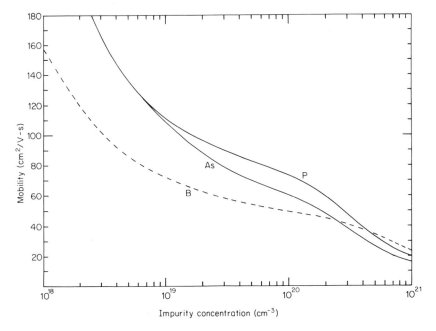

Figure 2.7. Measured values of majority carrier mobility as a function of impurity concentration in heavily doped silicon (after Masetti et al. [10], copyright © 1983 IEEE)

mobilities of arsenic and phosphorus doped material are significantly different [10], as illustrated in Figure 2.7. This indicates that lower sheet resistances can be achieved with phosphorus than arsenic.

In minority carrier devices such as bipolar transistors it is the minority carrier mobility that controls the electrical characteristics of the device. In the absence of information to the contrary it has usually been assumed that the minority and majority carrier mobilities are the same. However, recent measurements of minority carrier mobility suggest that this is not the case.

Figure 2.8 summarizes measured values of minority carrier hole mobility (i.e. hole mobility in n-type semiconductors) in silicon as a function of impurity concentration [11–14]. In the doping range 10^{17} to 10^{20} cm^{-3} the minority carrier mobility is over a factor of two higher than the equivalent majority carrier mobility. These experimental results are supported by theoretical calculations which predict that the minority carrier mobility is 2.8 times higher [15] at a doping concentration of around 5×10^{19} cm^{-3}. At very low impurity concentrations the two mobilities converge and are essentially identical. For the purposes of device modelling the minority carrier hole mobility can be described by the following empirical equation [11]:

$$\mu_p(\text{min}) = 130 + \frac{370}{1 + (N_d/8 \times 10^{17})^{1.25}} \text{ cm}^2/\text{V} - \text{s} \quad (2.52)$$

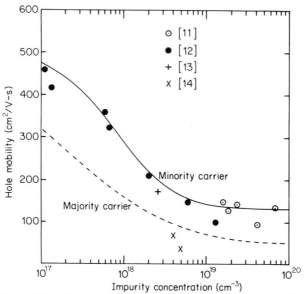

Figure 2.8. Measured values of minority carrier hole mobility as a function of impurity concentration in heavily doped silicon (after Del Alamo *et al.* [11], copyright ©1985 IEEE)

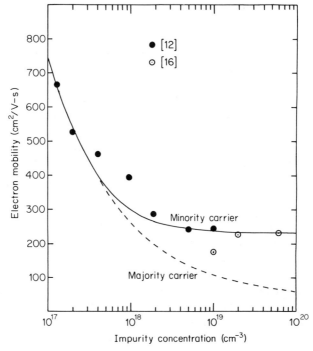

Figure 2.9. Measured values of minority carrier electron mobility as a function of impurity concentration in heavily doped silicon (after Swirhun *et al.* [16], copyright ©1986 IEEE)

The solid line in Figure 2.8 shows the fit between this equation and the experimental data.

Very few measurements of minority carrier electron mobility (i.e. electron mobility in *p*-type semiconductors) in silicon have been made, largely because *pnp* transistors are of less practical interest than *npn*. The most recent measured values [12,16] are illustrated in Figure 2.9 as a function of impurity concentration. In the doping range 10^{18} to 10^{20} cm^{-3} the minority carrier electron mobility decreases less strongly than the majority carrier one, and is a factor of more than two larger in heavily doped silicon. This factor of over two is considerably larger than the factor of 1.2 predicted by theory at a doping concentration around 5×10^{19} cm^{-3}. This discrepancy clearly demonstrates the need for further measurements to confirm these data. For the purposes of device modelling an empirical equation for the minority carrier electron mobility has been proposed [16]:

$$\mu_n(\min) = 232 + \frac{1180}{1 + \left(\dfrac{N_a}{8 \times 10^{16}}\right)^{0.9}} \text{ cm}^2/\text{V-s} \qquad (2.53)$$

The solid line in Figure 2.9 shows the fit between this equation and the experimental data.

2.7.2 Bandgap Narrowing

In lightly doped semiconductors the dopant atoms are sufficiently widely spaced in the semiconductor lattice that the wave functions associated with the dopant

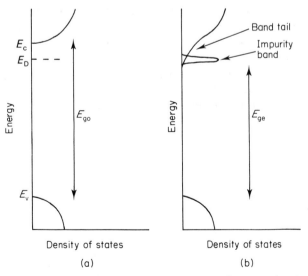

Figure 2.10. Energy versus density of states diagrams showing the effects of heavy doping in *n*-type silicon. (a) Lightly doped silicon; (b) heavily doped silicon

atoms' electrons do not overlap. The energy levels of the dopant atoms are therefore discrete. Furthermore, it is reasonable to assume that the widely spaced dopant atoms have no effect on the perfect periodicity of the semiconductor lattice, and hence the edges of the conduction and valence bands are sharply defined. This situation is illustrated in the density of states versus energy diagram in Figure 2.10(a).

In heavily doped semiconductors the dopant atoms are close enough together that the wave functions of their associated electrons overlap. This causes the discrete impurity level in Figure 2.10(a) to split and form an impurity band, as shown in Figure 2.10(b). In addition, the large concentration of dopant atoms disrupts the perfect periodicity of the silicon lattice, giving rise to a band tail instead of a sharply defined band edge. Figure 2.10(b) shows the density of states versus energy diagram for the case of a heavily doped, n-type semiconductor. It can be seen that the overall effect of the high dopant concentration is to reduce the bandgap from E_{go} to E_{ge}. A similar situation arises for a heavily doped p-type semiconductor, although in this case, of course, the bandgap narrowing occurs at the valence band edge.

The narrowing of the bandgap affects the pn product in the emitter, so that equation (2.30) is no longer valid. A convenient way of incorporating this into the simple theory is to modify the expression for the pn product:

$$p_{eo} n_{eo} = p_{eo} N_{de} = n_{ie}^2 = n_{io}^2 \exp \frac{\Delta E_{ge}}{KT} \quad (2.54)$$

where $\Delta E_{ge} = E_{go} - E_{ge}$ is termed the 'apparent bandgap narrowing in the emitter' and n_{io} the intrinsic carrier concentration for a lightly doped semiconductor. The relationship between the apparent and the actual bandgap narrowing is uncertain at present. Large discrepancies have been obtained between measured values of the silicon bandgap using absorption, photoluminescence and device measurements [17]. This serves to emphasize that the band structure of heavily doped silicon is not yet fully understood.

An alternative way of describing the effects of bandgap narrowing is through an effective doping concentration in the emitter N_{deff}:

$$N_{deff} = N_{de} \frac{n_{io}^2}{n_{ie}^2} = N_{de} \exp - \frac{\Delta E_{ge}}{KT} \quad (2.55)$$

This equation clearly indicates that the bandgap narrowing has the effect of reducing the effective doping concentration in the emitter, and hence also the gain of the bipolar transistor.

Measurement of the apparent bandgap narrowing in heavily doped n-type silicon is extremely difficult, and has been the subject of considerable controversy. However, Del Alamo et al. [17] have shown that the bandgap narrowing data in the literature are in remarkable agreement when they are interpreted in terms of the parameter N_{deff}/D_p. The inconsistencies in the

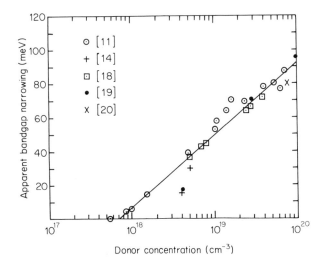

Figure 2.11. Measured values of apparent bandgap narrowing in n-type silicon as a function of donor concentration (after Del Alamo et al. [11], copyright ©1985 IEEE)

published data arise from difficulties in the measurement and interpretation of the diffusion coefficient D_p, or alternatively the measurement of the minority carrier mobility. Using the best available data for the minority carrier hole mobility (Figure 2.8), the apparent bandgap narrowing can be extracted from experimental results presented in the literature [17]. The results of this procedure are given in Figure 2.11, where the bandgap narrowing is shown as a function of donor concentration. It can be seen that bandgap narrowing starts to occur at a donor concentration of $7 \times 10^{17}\,\text{cm}^{-3}$, and then progressively increases as the donor concentration is increased. For the purposes of device modelling the apparent bandgap narrowing in the emitter ΔE_{ge} can be described by the following empirical equation [11]:

$$\Delta E_{ge} = 18.7 \ln\left(\frac{N_{de}}{7 \times 10^{17}}\right) \text{ meV} \tag{2.56}$$

The solid line in Figure 2.11 shows the fit between this equation and the experimental data.

The effective doping concentration in the emitter can be calculated from equation (2.55) using the apparent bandgap narrowing data in Figure 2.11, and is summarized in Figure 2.12. This figure clearly demonstrates how bandgap narrowing depresses the effective doping concentration in the emitter, and hence explains why the gains of practical silicon bipolar transistors are significantly lower than predicted by equation (2.45). Bandgap narrowing can be taken into account in equation (2.45) by replacing N_{de} with N_{deff}.

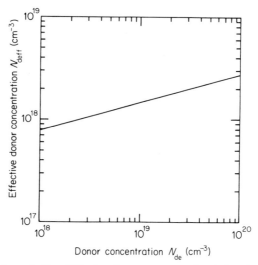

Figure 2.12. Effective donor concentration in heavily doped n-type silicon (after Del Alamo *et al.* [11], copyright ©1985 IEEE)

In general, bandgap narrowing can also occur in a heavily doped base, or indeed in the emitter of a *pnp* transistor. This can be dealt with in an analogous way by writing the *pn* product as:

$$n_{bo}p_{bo} = n_{bo}N_{ab} = n_{ib}^2 = n_{io}^2 \exp\frac{\Delta E_{gb}}{KT} \qquad (2.57)$$

where ΔE_{gb} is the apparent bandgap narrowing in the base.

Measurement of the apparent bandgap narrowing in heavily doped p-type silicon proceeds in a way similar to that described for n-type silicon. Using the best available minority carrier electron mobility data (Figure 2.9), the apparent bandgap narrowing can be extracted, as summarized in Figure 2.13. It can be seen that bandgap narrowing becomes important at an acceptor concentration of about 1×10^{17} cm^{-3}, and is significantly larger in p-type silicon than in n-type. These results indicate that bandgap narrowing occurs in the base regions of the great majority of practical silicon bipolar transistors. For the purposes of device modelling the apparent bandgap narrowing in the base ΔE_{gb} can be described by the following empirical equation [21]:

$$\Delta E_{gb} = 9\left[(F + (F^2 + 0.5)^{1/2}\right] \text{ meV} \qquad (2.58)$$

where

$$F = \ln\frac{N_a}{10^{17}} \qquad (2.59)$$

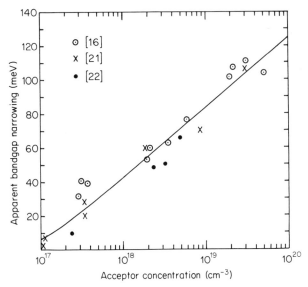

Figure 2.13. Measured values of apparent bandgap narrowing in p-type silicon as a function of acceptor concentration (after Swirhun et al. [16], copyright ©1986 IEEE)

The solid line in Figure 2.13 shows the fit between this equation and the experimental data.

Bandgap narrowing is also important in heavily doped gallium arsenide, as will be discussed in Chapter 5.

2.7.3 Auger Recombination

Experiments have shown [11,16] that the lifetime in heavily doped silicon is a strong function of doping concentration. These results can explained by the presence of an additional recombination mechanism which is important at high doping concentrations. Auger recombination [1,23] has been proposed as this mechanism. This is a three-particle, band-to-band recombination mechanism, in which the energy and momentum released by the recombination of an electron–hole pair is transferred to a free electron or hole. The Auger lifetime τ_A is given by:

$$\tau_A = \frac{1}{C_n N_{de}^2} \qquad (2.60)$$

where C_n is a constant, known as the Auger coefficient. The lifetime is therefore inversely proportional to the square of the donor concentration.

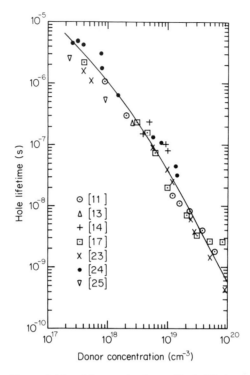

Figure 2.14. Measured values of hole lifetime in *n*-type silicon as a function of donor concentration (after Del Alamo et al. [11], copyright ©1985 IEEE)

Experimental values of hole lifetime in silicon as a function of donor concentration are summarized in Figure 2.14. The lifetime follows the dependence of equation (2.60) above a donor concentration of about 1×10^{19} cm^{-3}, indicating that Auger recombination is dominant in this regime. The experimental data can be fitted to an empirical equation of the form [11]:

$$\frac{1}{\tau_p} = 7.8 \times 10^{-13} N_d + 1.8 \times 10^{-31} N_d^2 \quad \text{s}^{-1} \tag{2.61}$$

as shown by the solid line in Figure 2.14. From equation (2.61) the Auger coefficient has a value of 1.8×10^{-31} cm^6/s. This is in good agreement with other measured values from the literature [28-30], which generally lie between 0.5 and 4.0×10^{-31} cm^6/s.

The hole diffusion length as a function of donor concentration can be calculated from equation (2.28) using the minority carrier hole mobility in Figure 2.8 and the hole lifetime in Figure 2.14. The results are shown by the solid line in Figure 2.15, and are compared with measured values of hole diffusion length from the literature.

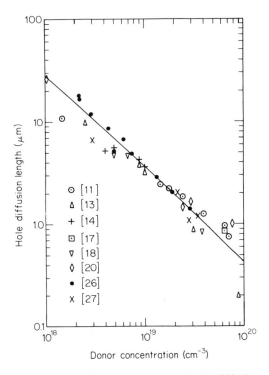

Figure 2.15. Measured values of hole diffusion length in *n*-type silicon as a function of donor concentration (after Del Alamo, Swirhun and Swanson [11], copyright ©1985 IEEE)

The electron lifetime varies with acceptor concentration in a way analogous to that described above for the hole lifetime. Unfortunately, measured values are relatively scarce, and also somewhat contradictory, as illustrated in Figure 2.16. Agreement between the two sets of data is reasonable at high acceptor concentrations but rather poor at lower ones. The lifetime follows the dependence of equation (2.60) at acceptor concentrations above about 5×10^{19} cm^{-3}, indicating that Auger recombination is dominant in this regime. For the purposes of device modelling an empirical expression has been proposed [16]:

$$\frac{1}{\tau_n} = 3.45 \times 10^{-12} N_a + 0.96 \times 10^{-31} N_a^2 \ \text{s}^{-1} \quad (2.62)$$

and is illustrated by the solid line in Figure 2.16. The Auger coefficient therefore takes a value of 0.95×10^{-31} cm^6/s.

The electron diffusion length can be calculated from equation (2.34) using the minority carrier electron mobility in Figure 2.9 and the electron lifetime

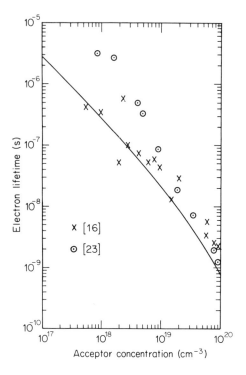

Figure 2.16. Measured values of electron lifetime in *p*-type silicon as a function of acceptor concentration (after Swirhun et al. [16], copyright © 1986 IEEE)

in Figure 2.16. The results are summarized in Figure 2.17 as a function of acceptor concentration.

2.8 EXTENSION OF THE SIMPLE THEORY

2.8.1 Heavy Doping Effects in Silicon Bipolar Transistors

The accurate prediction of the electrical characteristics of practical silicon bipolar transistors requires the incorporation of heavy doping effects into the simple theory. Unfortunately, this is, in general, not easy to achieve, particularly in transistors with non-uniform doping profiles. In this case all the parameters, such as lifetime, mobility and bandgap narrowing, vary with distance throughout the device. Exact computer simulations of the semiconductor equations is one possible approach to this problem. Although this is entirely effective, it tends to be expensive in computer time and also obscures the physics of the device behaviour. In this section we will therefore look for specific cases where meaningful analytical solutions are possible.

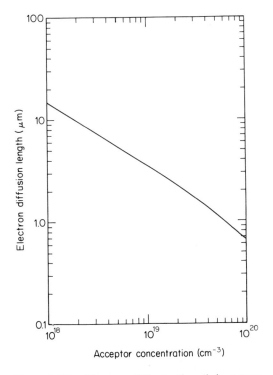

Figure 2.17. Electron diffusion length in p-type silicon as a function of acceptor concentration (after Swirhun et al. [16], copyright © 1986 IEEE)

In the bases of bipolar transistors the doping concentration is generally relatively low (typically 2×10^{18} cm^{-3}), and hence Auger recombination can, to a first approximation, be ignored. The spatial variation of doping concentration, bandgap narrowing and mobility can then be incorporated with only minor alterations into the basic equations. First, equation (2.46) can be modified by redefining the base Gummel number as:

$$G_b = \int_0^{W_B} \frac{N_{ab}(x)}{D_{nb}} \exp - \frac{\Delta E_{gb}}{KT} \, dx \tag{2.63}$$

where $N_{ab}(x)$, D_{nb} and ΔE_{gb} all vary with distance through the base. The equation for the collector current can then be written as:

$$I_C = - \frac{qAn_{io}^2}{G_b} \exp \frac{qV_{BE}}{KT} \tag{2.64}$$

where n_{io} is the intrinsic carrier concentration of lightly doped silicon. This is related to the intrinsic carrier concentration of heavily doped p-type silicon n_{ib} through equation (2.57).

In the emitter of the bipolar transistor the situation is more complicated, because the doping concentration is high enough for Auger recombination to be important. Furthermore, in many cases the emitter doping is sufficiently high that the majority carriers are degenerate. In this case, Fermi–Dirac statistics must be used. These two additional factors mean that the simple theory in Section 2.4 is no longer valid, and hence a more rigorous analysis is required. Unfortunately, there is no simple means of obtaining a general analytical solution to the semiconductor equations under these conditions. However, simple and reasonably accurate solutions can be obtained in some specific types of emitter.

One group of emitters for which an analytical solution is available is shallow emitters, as described in Section 2.6. These are emitters where their depth is small with respect to the minority carrier diffusion length, so that negligible recombination occurs in the emitter. The base current for this case has been derived by Shibib et al. [32]:

$$I_B = \frac{qAn_{io}^2}{\int_0^{W_E} (N_{\text{deff}}(x)/D_{pe})\,dx + (N_{\text{deff}}(W_E)/S_M)} \exp \frac{qV_{BE}}{KT} \quad (2.65)$$

where N_{deff} is given by equation (2.55) and is a function of distance through the emitter, $N_{\text{deff}}(W_E)$ is the value of the effective doping concentration at the surface of the emitter and S_M is the surface recombination velocity. For a shallow emitter with a metal contact the recombination velocity is large, and equation (2.65) can be simplified to:

$$I_B = \frac{qAn_{io}^2}{G_e(W_E)} \exp \frac{qV_{BE}}{KT} \quad (2.66)$$

where

$$G_e(W_E) = \int_0^{W_E} \frac{N_{\text{deff}}(x)}{D_{pe}}\,dx = \int_0^{W_E} \frac{N_{de}(x)}{D_{pe}} \exp - \frac{\Delta E_{ge}}{KT}\,dx \quad (2.67)$$

$G_e(W_E)$ is the emitter Gummel number of the transistor, which is defined in a way analogous to the base Gummel number in equation (2.63). The only unknown in equation (2.67) is the emitter doping profile $N_{de}(x)$, since the diffusion coefficient D_{pe} and bandgap narrowing ΔE_{ge} can be obtained from Figures 2.8 and 2.11. The base current for this type of emitter can therefore be calculated from equation (2.66) using simple numerical integration routines.

Other authors [33] have used a regional analysis to derive an expression for the base current of a shallow emitter. This assumes that the neutral emitter can be divided into two regions, as illustrated in Figure 2.18. All recombination occurs in the heavily doped region close to the surface, and the recombination mechanism is assumed to be Auger. The heavy doping effects discussed in Section 2.7 are important in this region, and are all modelled. The lightly doped

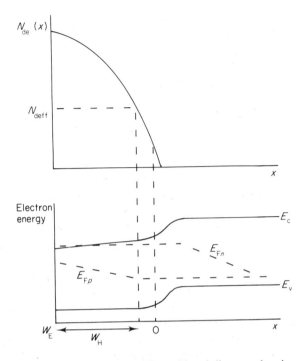

Figure 2.18. Doping profile and band diagram showing the division of an emitter into lightly and heavily doped regions (after Fossum and Shibib [33], copyright ©1981 IEEE)

region close to the emitter/base junction is transparent to minority carriers and acts as a buffer layer between the heavily doped region and the moving edge of the emitter/base depletion region. This partitioning of the neutral emitter simplifies the solution of the transport equations and allows the following simple equation for the base current to be derived [33]:

$$I_B = \frac{qAn_{io}^2}{N_{deff}} \left(\frac{D_{pe}(W_E)}{W_H} \right) \exp \frac{qV_{BE}}{KT} \tag{2.68}$$

where $D_{pe}(W_E)$ is the hole diffusion coefficient at the surface and N_{deff} is defined by equation (2.55). Figure 2.12 shows that the effective doping concentration N_{deff} is approximately constant for doping densities in the range $10^{18} < N_{de} < 10^{20}$ cm^{-3}, with a value of around 1×10^{18} cm^{-3}. W_H is defined in Figure 2.18, and is essentially the distance below the surface at which bandgap narrowing becomes important, i.e. where $N_{de} = N_{deff}$.

Equations (2.65)–(2.68) are valid provided that the emitter is shallow enough that the majority of recombination occurs at the surface. As the emitter depth is increased, however, an increasing fraction of minority carriers recombine in the bulk emitter, and these equations become progressively more inaccurate.

A first-order correction for recombination in the emitter can be derived by modifying equation (2.65) as follows [34]:

$$I_B = \frac{qAn_{io}^2}{G_e(W_E) + (N_{deff}(W_E)/S_M)} \left[1 + \int_0^{W_E} \frac{[G_E(W_E) - G_e(x)]}{\tau_{pe}(x)N_{deff}(x)} dx \right.$$

$$\left. + \frac{N_{deff}(W_E)}{S_M} \int_0^{W_E} \frac{dx}{\tau_{pe}(x)N_{deff}(x)} \right] \exp \frac{qV_{BE}}{KT} \quad (2.69)$$

where τ_{pe} represents the spatial variation of the lifetime through the emitter and the term in square brackets is the correction for recombination in the emitter. This quasi-empirical equation for the base current can be solved using numerical integration routines, and is reasonably accurate for emitter/base junction depths of less than about 0.3 μm. The majority of high-speed bipolar transistors have emitter/base junction depths of less than 0.3 μm, and hence equation (2.69) is applicable to this type of device.

In power bipolar transistors the emitters are designed to be very deep in order to produce the necessary high breakdown voltages. This type of emitter is essentially opaque to minority carriers, and hence equation (2.69) is inappropriate. However, it has been shown [33] that equation (2.68) provides a reasonable approximation for the base current in these circumstances. Differences in base current in power transistors can therefore be explained by differences in the width of the heavily doped surface region W_H in Figure 2.18.

2.8.2 Low-current Gain

The simple theory in the previous sections predicts that both the collector and base currents vary as $\exp(qV_{BE}/KT)$, and hence that the gain is constant. However, in practical silicon [2] and heterojunction [35] transistors the gain is constant over only a very small range of collector currents, and decreases markedly at both low and high currents. This is illustrated in Figure 2.19 for a practical silicon bipolar transistor. Although the collector current varies in the expected way, the base current at low forward voltages shows an exp (qV_{BE}/mKT) dependence, with the ideality factor m taking a value between 1 and 2. In this section it will be shown that these experimental results can be explained by recombination in the emitter/base depletion region [36], which was ignored in the simple theory (see Section 2.3.1). We will begin by considering the physics of the recombination process, and proceed to show how the simple theory can be modified to take into account recombination in the depletion region.

Figure 2.19 *(opposite)*. (a) Common emitter current gain as a function of collector current for a practical silicon bipolar transistor; (b) Gummel plot for a practical silicon bipolar transistor showing the presence of a component of base current due to recombination in the emitter/base depletion region

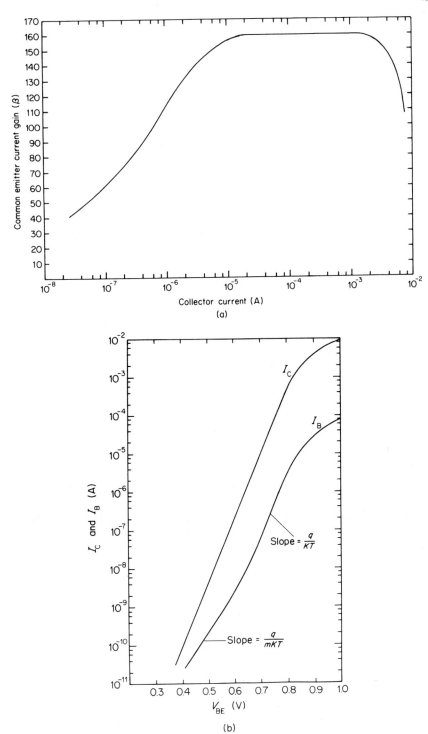

Recombination via deep levels

Recombination in wide bandgap semiconductors such as silicon and gallium arsenide generally occurs via deep levels located close to the centre of the bandgap. These deep levels or recombination centres arise from imperfections and impurities, and have the effect of disrupting the perfect periodicity of the semiconductor lattice. They thereby give rise to discrete energy levels in the bandgap in a similar way to donor and acceptor levels. This type of recombination is very efficient, because the deep levels act as 'stepping stones', aiding the transition of electrons and holes between the conduction and valence bands.

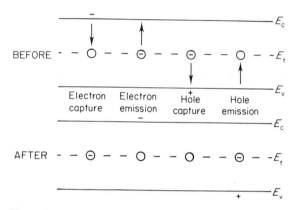

Figure 2.20. Generation/recombination processes via a deep level in the bandgap

Figure 2.20 shows the four transitions that can occur when recombination occurs via a single deep level. The first transition is electron capture, where an electron drops from the conduction band into the deep level. The rate of electron capture r_{nc} is proportional to the number of electrons in the conduction band n and the number of deep levels which are not occupied by electrons p_t:

$$r_{nc} = c_n n p_t \tag{2.70}$$

The proportionality constant can be written as $c_n = v_{th} \sigma_n$, where v_{th} is the thermal velocity:

$$v_{th} = \left[\frac{3KT}{m^*} \right]^{1/2} \tag{2.71}$$

The quantity σ_n is the capture cross-section for electrons and is a measure of how close the electron must come to the deep level to be captured.

The second transition in Figure 2.20 is electron emission from the deep level into the condution band. The rate of electron emission r_{ne} is proportional to the number of deep levels that are occupied by electrons n_t:

$$r_{ne} = e_n n_t \tag{2.72}$$

The proportionality constant e_n is the electron emission probability, which can be calculated from the equilibrium case, where the net recombination rate $(r_{nc} - r_{ne})$ is equal to zero. In equilibrium, the electron concentration is given by equation (1.1), and the number of deep levels occupied by electrons can be written as [2]:

$$n_t = \frac{N_t}{1 + \exp(E_t - E_F)/KT} \tag{2.73}$$

where n_t is the total number of deep levels and E_t its energy level in the bandgap. The emission probability can therefore be calculated from equations (2.70)–(2.73):

$$e_n = v_{th} \sigma_n n_i \exp\left(\frac{E_t - E_i}{KT}\right) \tag{2.74}$$

This equation shows that the emission probability increases exponentially as the energy level E_t moves away from the centre of the bandgap towards the conduction band. This is intuitively what we would expect, since levels close to the conduction band have a high probability of being empty of electrons.

An analogous set of equations can also be derived for hole capture and emission, namely:

$$r_{pc} = v_{th} \sigma_p p n_t \tag{2.75}$$

$$r_{pe} = p_t v_{th} \sigma_p n_i \exp\left(\frac{E_i - E_t}{KT}\right) \tag{2.76}$$

Under steady-state, non-equilibrium conditions, such as are found in a forward-biased emitter/base junction, the rate at which electrons enter the conduction band must equal that at which they leave. Similarly, the rate at which holes enter the valence band must equal that at which they leave, and the net generation rate G must equal the net recombination rate U. We can therefore write:

$$U = r_{nc} - r_{ne} = r_{pc} - r_{pe} \tag{2.77}$$

giving

$$U = \frac{\sigma_n \sigma_p v_{th} N_t (pn - n_i^2)}{\sigma_n \{n + n_i \exp[(E_t - E_i)/KT]\} + \sigma_p \{p + n_i \exp[(E_i - E_t)/KT]\}} \tag{2.78}$$

Recombination current in the emitter/base depletion region

The Gummel plot in Figure 2.19 shows that there are two distinct regions to the base characteristic, which are identified by slopes of q/mKT and q/KT. The q/KT slope at high currents indicates that the diffusion current (Section 2.4) is dominant in this portion of the characteristic, while the q/mKT slope at low currents is indicative of recombination in the emitter/base depletion region. These results imply that the recombination current in the emitter/base depletion region can be treated as an independent component of base current and added to the other components to give the total base current. Numerical simulations [37] confirm that this assumption is reasonable, and indicate that the base current can be accurately described by an equation of the form:

$$I_B = I_{pe} + I_{rg} = I_1 \exp \frac{qV_{BE}}{KT} + I_2 \exp \frac{qV_{BE}}{mKT} \tag{2.79}$$

The first term models the diffusion current given by equation (2.29) and the second term the recombination current in the emitter/base depletion region which will now be derived.

In a forward-biased emitter/base junction the electrons lost by recombination in the depletion region give rise to a recombination current:

$$I_{rg} = qA \int_0^{W_D} U \, dx \tag{2.80}$$

where W_D is the emitter/base depletion width. Unfortunately, this equation is very difficult to solve analytically in forward bias, because the recombination rate U is a function of the electron and hole concentrations, which vary with distance across the depletion region. In this section we will take the approach of simplifying the equation for the recombination rate, and hence derive an approximate analytical equation for the recombination current.

As explained above, the most effective recombination centres lie at the centre of the bandgap. It will therefore be assumed that all the recombination centres have an energy of $E_t = E_i$. This is the condition for maximum recombination, and hence can be considered as the worst case. A considerable simplification of equation (2.78) results if it is also assumed that $\sigma_n = \sigma_p = \sigma$, so that:

$$U = \sigma v_{th} N_t \frac{pn - n_i^2}{n + p + 2n_i} \tag{2.81}$$

From the assumption of quasi-equilibrium (Section 2.4.1) the non-equilibrium pn product is a constant, given by:

$$pn = n_i^2 \exp \frac{qV_{BE}}{KT} \tag{2.82}$$

The recombination rate U is therefore:

$$U = \frac{\sigma v_{\text{th}} N_t n_i^2}{n+p+2n_i}\left(\exp\frac{qV_{\text{BE}}}{KT} - 1\right) \quad (2.83)$$

The recombination rate U is a maximum when the sum of the carrier concentrations $(p+n)$ is a minimum. It can easily be shown that the condition for a minimum is $p = n$, which occurs at the point in the depletion region where the intrinsic Fermi level is mid-way between the electron quasi-Fermi level and the hole quasi-Fermi level. Equation (2.82) shows that at this point the carrier concentrations are given by:

$$p = n = n_i \exp\frac{qV_{\text{BE}}}{2KT} \quad (2.84)$$

Substituting into equation (2.83) then gives the maximum recombination rate:

$$U_{\max} \simeq \frac{1}{2}\sigma v_{\text{th}} N_t n_i \exp\frac{qV_{\text{BE}}}{2KT} \quad (2.85)$$

where it has been assumed that $V_{\text{BE}} \gg KT/q$.

An estimate for the recombination current in the emitter/base depletion region can be obtained if it is assumed that the recombination rate U is equal its maximum value $U_{\max}$ throughout the depletion region. Evaluation of the integral in equation (2.80) is then straightforward:

$$I_{\text{rg}} = \frac{1}{2}qAW_D \sigma v_{\text{th}} N_t n_i \exp\frac{qV_{\text{BE}}}{2KT} \quad (2.86)$$

This equation essentially gives the recombination current for the case of strong recombination, and predicts an $\exp(qV_{\text{BE}}/2KT)$ dependence. This type of behaviour is often seen in practical GaAs/GaAlAs heterojunction transistors [35], since lattice mismatch at the interface between the GaAs and GaAlAs tends to produce a large density of recombination centres in the emitter/base depletion region. Equation (2.86) also predicts that the recombination current increases with the density of traps N_t and the capture cross-section σ, which is intuitively what would be expected.

In practical silicon transistors it is more usual for the recombination current I_{rg} to follow an $\exp(qV_{\text{BE}}/mKT)$ dependence, with the ideality factor m taking a value between 1 and 2. For example, the base characteristic in Figure 2.19 has an ideality factor of approximately 1.7 at low currents. This type of behaviour has been predicted in the literature [37], where exact numerical solutions of equation (2.80) have been obtained. The precise value of m has been shown to depend upon the physical properties of the deep level. This is

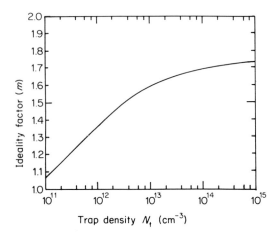

Figure 2.21. Theoretical variation of the ideality factor *m* with trap density (after Ashburn *et al.* [37], copyright ©1975 Pergamon Journals Ltd)

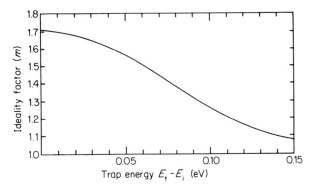

Figure 2.22. Theoretical variation of the ideality factor *m* with trap energy (after Ashburn *et al.* [37], copyright ©1975 Pergamon Journals Ltd)

illustrated in Figures 2.21 and 2.22, where the ideality factor is plotted as a function of trap density and position in the bandgap. These predictions show that the ideality factor increases with trap density and decreases as the trap moves away from band centre.

2.8.3 High-current Gain

In the simple theory of Section 2.4 it was assumed that the number of electrons injected from the emitter into the base was small with respect to the doping concentration in the base. This assumption is reasonable at moderate current levels, but at high currents the injected electron concentration may become much greater than the base doping concentration. When this occurs the hole concentration in the base must increase by the same amount as the electron concentration in order to maintain charge neutrality. This regime of transistor operation is referred to as conductivity modulation or high-level injection. Effectively, the base becomes more heavily doped as the injection level increases. Since equation (2.43) shows that the collector current is inversely proportional to the base doping concentration, this has the effect of reducing the rate of increase of the collector current. A rigorous analysis [38] shows that the collector current varies as:

$$I_C = I_3 \exp \frac{qV_{BE}}{2KT} \qquad (2.87)$$

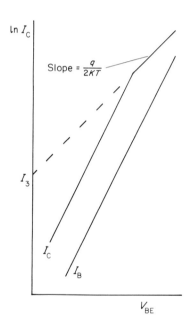

Figure 2.23. Gummel plot showing high-level injection effects in a bipolar transistor

Figure 2.23 illustrates the predicted behaviour, from which it can be seen that the slope of the collector characteristic changes from q/KT to $q/2KT$ on entering high-level injection. This leads to a decrease in gain at high currents.

High-level injection is particularly important in devices in which the base doping is very low. For example, power devices such as PIN diodes or thyristors often exhibit a clearly defined high-level injection region, beginning at forward biases as low as 0.4 V [39]. In most bipolar transistors, however, the base doping is relatively high and a clearly defined transition from low- to high-level injection is rarely discernable. For example, the turnover of the collector characteristic at high currents in Figure 2.19 is only partly due to high-level injection effects.

Series collector and emitter resistance are also important at these current levels, and these tend to mask the transition to high-level injection.

2.8.4 Basewidth Modulation

The simple theory in Section 2.4 gives no indication of how the current gain is affected by the collector/base voltage. The function of the collector/base junction in a bipolar transistor is merely to gather the minority carriers injected from the emitter into the base. We would therefore expect the collector/base voltage to have very little effect on the gain, provided, of course, that the junction does not become forward biased. Although this is true in the majority of transistors, when the base doping is low or the basewidth very small, the application of a reverse bias to the collector can cause the width of the base to be significantly modulated.

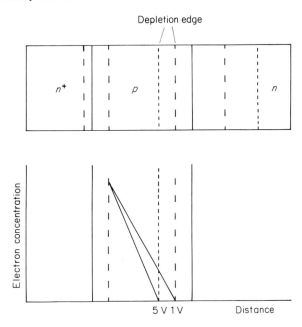

Figure 2.24. Minority carrier distribution in the base at two values of collector/base reverse voltage

Figure 2.24 illustrates the effect of an increase in the collector/base voltage on the depletion width and the minority carrier electron distribution in the base. In a heavily doped base the majority of this increase in depletion width occurs in the collector, but in a lightly doped one significant penetration into the base can occur. This leads to a narrowing of the neutral basewidth, and a consequent increase in the gradient of the injected electron distribution in the base. Equation (2.16) shows that the electron diffusion current is proportional to this gradient, and hence it is clear that an increase in collector/base reverse bias leads directly to an increase in collector current.

Figure 2.4 shows Gummel plots for a practical high-gain bipolar transistor under collector/base reverse biases of 1 V and 5 V. A 15% increase in collector current is obtained, indicating that significant modulation of the basewidth is occurring. Similarly, Figure 2.25 shows the effect of basewidth modulation on transistor output characteristics. Here, it gives rise to a finite slope on the characteristics, which is equivalent to a conductance at the output of the transistor. It is also interesting to note that if the individual characteristics in Figure 2.25 are extrapolated back along the voltage axis they originate from a single point. This extrapolated voltage is known as the Early voltage V_{AF}, and is used as a model parameter in computer-aided circuit design programs (Chapter 3).

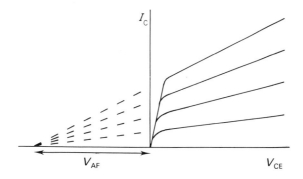

Figure 2.25. Bipolar transistor output characteristic showing the effect of basewidth modulation

2.9 JUNCTION BREAKDOWN

The transistor characteristic in Figures 1.8(a) and 1.9 indicate that a large current flows through the device at high values of collector/base voltage. The voltage at which this increase in current occurs is known as the breakdown voltage. It is clear from Figures 1.8 and 1.9 that no transistor action is obtained above this voltage, and hence that it imposes an upper limit on the operating voltage of the device. It is also interesting to note that a lower breakdown voltage is obtained when the transistor is connected in the common emitter mode than in the common base one. At first sight, this is somewhat surprising, since in both cases it is the collector/base junction that is breaking down. In this section we will explain how the current gain of the transistor is responsible for this difference.

Several physical mechanisms can give rise to excessive current at high collector voltages, the most important of which are punch-through, zener breakdown and avalanche breakdown. The first two mechanisms can usually be avoided by careful transistor design, but avalanche breakdown imposes a fundamental limit on the operating voltage of bipolar transistors.

2.9.1 Punch-through

In Section 2.8.4 it was shown how the application of a reverse bias to the collector caused the collector/base depletion region to extend into the base and hence modulate the basewidth. In transistors with very narrow or very lightly doped bases the application of a reverse bias to the collector can cause the depletion region to extend across the whole width of the base and join up with the emitter/base depletion region. The emitter and collector are then connected together by a single depletion region, as illustrated in Figure 2.26. This is known as punch-through, and when it occurs a large current flows between emitter and collector. Its electrical effect is similar to junction breakdown, although, of course, the physical mechanism is completely different.

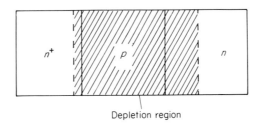

Figure 2.26. Schematic illustration of a bipolar transistor operating in punch-through

State-of-the-art, high-speed bipolar transistors typically have basewidths of less than 0.1 μm, and consequently often operate close to the punch-through limit. Carful transistor and process design is therefore required in order to ensure that punch-through does not degrade the yield of the process. From these considerations it is also clear that punch-through imposes a fundamental limit to the scaling of the basewidth of a bipolar transistor.

2.9.2 Zener Breakdown

Zener breakdown [1,2] is a tunnelling mechanism in which large numbers of electrons penetrate through the energy barrier imposed by the bandgap of the semiconductor. This is illustrated schematically in Figure 2.27 for a reverse-biased *pn* junction. In order for tunnelling to occur, the barrier presented to the tunnelling electrons must be very thin. This situation only arises at electric fields above approximately 10^6 V/cm. In general, such high electric fields only occur when both the *n* and *p* regions are very heavily doped. In most practical transistors, therefore, tunnelling does not take place, and only becomes a problem at high base doping concentrations [40].

Zener breakdown can be distinguished from avalanche breakdown by measuring the temperature dependence of the breakdown voltage. If Zener

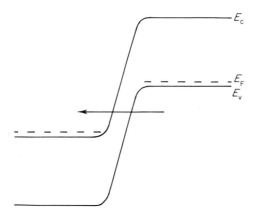

Figure 2.27. Band diagram illustrating Zener breakdown

breakdown is the dominant mechanism, the breakdown voltage decreases with increasing temperature, whereas for avalanche breakdown it increases [41].

2.9.3 Avalanche Breakdown

Avalanche multiplication or impact ionization [1,2] is by far the most common breakdown mechanism in practical bipolar transistors. In a reverse-biased *pn* junction electron–hole pairs are continually being generated by thermal agitation. At low reverse voltages this gives rise to a leakage generation current, which can easily be calculated from equation (2.80) [2]. At high reverse voltages, however, the generated carriers gain sufficient kinetic energy between collisions with the silicon lattice for them to be able to shatter the silicon–silicon bond. This mechanism is referred to as impact ionization, and leads to the generation of an electron–hole pair. The original carrier and the electron–hole pair are then accelerated by the electric field, and in turn are able to produce further electron–hole pairs by impact ionization. This process, known as avalanche multiplication, rapidly leads to the generation of a large number of carriers and hence to a large reverse current.

In order for avalanche multiplication to occur, a critical electric field E_{crit} must be established across the reverse-biased junction. Since the depletion width depends upon the doping concentration it is clear that the breakdown voltage BV will also depend upon the doping concentration. For a one-sided step junction the breakdown voltage is given by [1]:

$$BV = \frac{\varepsilon_0 \varepsilon_r E_{\text{crit}}^2}{2 q N_L} \qquad (2.88)$$

where N_L is the doping concentration on the lightly doped side of the junction. If E_{crit} was a constant, equation (2.88) would indicate that the breakdown voltage was inversely proportional to the doping concentration. In practice, however, E_{crit} varies slightly with doping concentration [1], taking values between 3×10^5 and 1×10^6 V/cm for silicon and gallium arsenide [1,42].

In bipolar transistors the collector/base breakdown voltage with the base open circuit BV_{CBO} is given by equation (2.88), in which N_L is replaced by the collector doping concentration. However, in the common emitter mode (Figure 1.8) the breakdown voltage BV_{CEO} is considerably lower than BV_{CBO}. This can be understood by considering the currents flowing in the transistor when the base is open-circuit. The application of a positive voltage to the collector and a negative voltage to the emitter (Figure 1.7(a)) causes not only the collector/base junction to be reverse biased but also the emitter/base junction to be slightly forward biased. The collector current is given by the sum of a generation leakage current I_g and a current resulting from the slightly forward biased emitter/base junction. Since the base is open-circuit, the collector current must also be equal to the emitter current, giving:

$$I_C = I_E = \alpha I_E + I_g \qquad (2.89)$$

The generation current I_g is merely the leakage current of the collector/base junction I_{CBO}. Equation (2.89) can therefore be written as:

$$I_{CEO} = \alpha I_{CEO} + I_{CBO} \qquad (2.90)$$

where I_{CEO} is the leakage current obtained between collector and emitter when the base is left open-circuit.

When the collector/base reverse voltage is high enough for avalanche multiplication to occur, the collector/emitter leakage current is greatly increased. This can be taken into account by defining a multiplication factor M, such that:

$$I_{CEO} = (\alpha I_{CEO} + I_{CBO})M \qquad (2.91)$$

giving

$$I_{CEO} = \frac{M}{1 - \alpha M} I_{CBO} \qquad (2.92)$$

Equation (2.92) shows that the collector/emitter current begins to increase very rapidly when αM approaches unity. In contrast, in the common base mode (Figure 1.9) the collector/base leakage current only begins to increase when αM approaches infinity. This clearly explains why the breakdown voltage in the common emitter mode BV_{CEO} is lower than that in the common base mode BV_{CBO}.

The magnitude of BV_{CEO} can be calculated by using an empirical expression [43] for the multiplication factor:

$$M = \left[1 - \left(\frac{V_{CB}}{BV_{CBO}}\right)^n\right]^{-1} \qquad (2.93)$$

where n takes values between 3 and 6. Breakdown in the common emitter mode occurs when $V_{CB} = BV_{CEO}$ and $\alpha M = 1$, giving:

$$BV_{CEO} = BV_{CBO} \sqrt[n]{(1-\alpha)} \simeq \frac{BV_{CBO}}{\sqrt[n]{\beta}} \qquad (2.94)$$

The common emitter breakdown voltage BV_{CEO} is therefore inversely proportional to the common emitter current gain of the transistor, indicating that high gain and high breakdown voltage cannot be obtained simultaneously in a bipolar transistor.

References

1. S. M. Sze, *Physics of Semiconductor Devices*, John Wiley, Chichester (1985).
2. A. S. Grove, *Physics and Technology of Semiconductor Devices*, John Wiley, Chichester (1971).
3. H. J. de Man, 'The influence of heavy doping on the emitter efficiency of a bipolar transistor', *IEEE Trans. Electron. Devices*, **ED18**, 833 (1971).
4. W. L. Kauffman and A. A. Bergh, 'The temperature dependence of ideal gain in double diffused silicon transistors', *IEEE Trans. Electron. Devices*, **ED15**, 732 (1968).
5. S. M. Sze and J. C. Irvin, 'Resistivity, mobility, and impurity levels in GaAs, Ge, and Si at 300 K, *Solid State Electronics*, **11**, 599 (1968).
6. R. J. Van Overstraeten, H. J. De Man and R. P. Mertens, 'Transport equations in heavily doped silicon', *IEEE Trans. Electron. Devices*, **ED20**, 290 (1973).
7. J. S. Blakemore, *Semiconductor Statistics*, Pergamon Press, Oxford (1962).
8. G. L. Pearson and J. Bardeen, 'Electrical properties of pure silicon and silicon alloys containing boron and phosphorus', *Phys. Rev.*, **75**, 865 (1949).
9. W. R. Thurber, R. L. Mattis, Y. M. Liu and J. J. Filliben, 'Resistivity–dopant density relationship for boron-doped silicon', *Jnl Electrochem. Soc.* **127**, 2291 (1980).
10. G. Masetti, M. Severi and S. Solmi, 'Modeling of carrier mobility against carrier concentration in arsenic, phosphorus, and boron doped silicon', *IEEE Trans. Electron. Devices*, **ED30**, 764 (1983).
11. J. del Alamo, S. Swirhun and R. M. Swanson, 'Simultaneous measurement of hole lifetime, hole mobility, and bandgap narrowing in heavily doped n-type silicon', *IEDM Technical Digest*, 290 (1985).
12. J. Dziewor and D. Silber, 'Minority carrier diffusion coefficients in highly-doped silicon', *App. Phys. Lett.*, **35**, 170 (1979).
13. D. E. Burk and V. de la Torre, 'An empirical fit to minority hole mobilities', *IEEE Electron. Device Lett.*, **EDL5**, 231 (1984).
14. R. Mertens, J. Van Meerbergen, J. Nijs and R. Van Overstraeten, 'Measurement of the minority carrier transport parameters in heavily doped silicon', *IEEE Trans. Electron. Devices*, 27, 949 (1980).
15. H. S. Bennett, 'Hole and electron mobilities in heavily doped silicon: comparison of theory and experiment', *Solid State Electronics*, **26**, 1157 (1983).

16. S. E. Swirhun, Y. H. Kwark and R. M. Swanson, 'Measurement of electron lifetime, electron mobility, and bandgap narrowing in heavily doped p-type silicon', *IEDM Technical Digest*, 24 (1986).
17. J. Del Alamo, S. Swirhun and R. M. Swanson, 'Measuring and modeling minority carrier transport in heavily doped silicon', *Solid State Electronics*, **28**, 47 (1985).
18. A. Wieder, 'Emitter effects in shallow bipolar devices', *IEEE Trans. Electron. Devices*, **ED27**, 1402 (1980).
19. A. Neugroschel, S. C. Pao and F. A. Lindholm, 'A method for determining energy gap narrowing in highly doped semiconductors', *IEEE Trans. Electron. Devices*, **ED29**, 894 (1982).
20. G. E. Possin, M. S. Adler and B. J. Baliga, 'Measurement of the pn product in heavily doped epitaxial emitters', *IEEE Trans. Electron. Devices*, **ED31**, 3 (1984).
21. J. W. Slotboom and H. C. de Graaff, 'Measurement of bandgap narrowing in silicon bipolar transistors', *Solid State Electronics*, **19**, 857 (1976).
22. M. Ghannam, PhD thesis, Katholieke Universiteit, Leuven (1985).
23. J. Dziewior and W. Schmid, 'Auger coefficients for highly doped and highly excited silicon', *App. Phys. Lett.*, **31**, 346 (1977).
24. J. D. Beck and R. Conradt, 'Auger recombination in silicon', *Solid State Comm.*, **13**, 93 (1977).
25. Y. Valtkus and V. Grivitskas, 'Dependence of the rate of Auger recombination on the carrier density in silicon', *Sov. Phys. Semicond.*, **15**, 1102 (1981).
26. L. Pasari and E. Susi, 'Recombination mechanisms and doping density in silicon', *Jnl App. Phys.*, **54**, 3935 (1983).
27. J. F. Nijs, PhD thesis, Katholieke Universiteit, Leuven (1982).
28. A. Haug, 'Carrier density dependence of Auger recombination', *Solid State Electronics*, **21**, 1281 (1978).
29. D. J. Roulston, N. D. Arora and S. G. Chamberlain, 'Modelling and measurement of minority carrier lifetime in heavy doped N diffused silicon diodes', *IEEE Trans. Electron. Devices*, **ED29**, 284 (1982).
30. J. Dziewior and W. Schmid, 'Auger coefficients for highly doped and highly excited silicon', *App. Phys. Lett.*, **31** 346 (1977).
31. K. G. Svantesson and N. G. Nilson, 'Measurement of Auger recombination in silicon by laser excitation', *Solid State Electronics*, **21**, 1603 (1978).
32. M. A. Shibib, F. A. Lindholm and F. Therez, 'Heavily doped transparent emitter regions in junction solar cells, diodes and transistors', *IEEE Trans. Electron. Devices*, **ED26**, 959 (1979).
33. J. G. Fossum and M. A. Shibib, 'An analytic model for minority carrier transport in heavily doped regions of silicon devices', *IEEE Trans. Electron. Devices*, **ED28**, 1018 (1981).
34. J. A. Del Alamo and R. M. Swanson, 'The physics and modeling of heavily doped emitters', *IEEE Trans. Electron. Devices*, **ED31**, 1878 (1984).
35. H. Ito, 'Generation–recombination current in the emitter/base junction of AlGaAs/GaAs HBTs', *Japan Jnl App. Phys.*, **25**, 1400 (1986).
36. C. T. Sah, R. N. Noyce and W. Shockley, 'Carrier generation and recombination in pn junctions and pn junction characteristics', *Proc. IRE*, **45**, 1228 (1957).
37. P. Ashburn, D. V. Morgan and M. J. Howes, 'A theoretical and experimental study of recombination in silicon p–n junctions', *Solid State Electronics*, **18**, 569 (1975).
38. R. N. Hall, 'Power rectifiers and transistors', *Proc. IRE*, **44**, 72 (1956).
39. B. W. Wessels and B. J. Baliga, 'Vertical channel field controlled thyristors with high gain and fast switching speeds', *IEEE Trans. Electron. Devices*, **ED25**, 1261 (1978).

40. A. Cuthbertson and P. Ashburn, 'Self-aligned transistors with polysilicon emitters for bipolar VLSI', *IEEE Trans. Electron. Devices*, **ED32**, 242 (1985).
41. M. J. O. Strutt, *Semiconductor Devices*, Vol 1, Academic Press, London, (1966).
42. S. M. Sze and G. Gibbons, 'Avalanche breakdown voltages of abrupt and linearly graded *pn* junctions in Ge, Si, GaAs, and GaP', *App. Phys. Lett.*, **8**, 111 (1966).
43. S. L. Miller, 'Ionization rates for electrons and holes in silicon', *Phys. Rev.*, **99**, 1234 (1955).

Chapter 3
BIPOLAR TRANSISTOR MODELS

3.1 TRANSISTOR MODELLING

The efficient design of integrated circuits increasingly requires the use of sophisticated computer-aided circuit design programs such as the ubiquitous SPICE program. These take a circuit description as input and provide output in the form of node voltages and currents as a function of time. A vital component of these programs is transistor models, which define the terminal characteristics of the active devices. These models consist of a combination of circuit elements such as resistors, capacitors, current generators, etc; and equations for defining the non-linear behaviour of the transistor. Input to the transistor models is through a set of transistor parameters, typically 40 for a full description of a bipolar transistor.

In devising a transistor model for inclusion in a circuit design program, an accurate description of the terminal characteristics is more important than a rigorous description of the device physics. Nevertheless, models that are based on the physics of the device do provide a better understanding, and can generally be implemented using fewer model parameters. For these reasons, most computer-aided circuit design programs use transistor models that are physics based, although simple empirical expressions are often used to describe aspects of device behaviour. Computational time is also of paramount importance, since this provides a limit to the size of circuit that can be simulated. Clearly, there is therefore a trade-off between accuracy and complexity.

Transistor models provide an interface between process engineers, device engineers and integrated circuit designers. Circuit designers need to be familiar with transistor models, because the accuracy of their circuit simulations depend to a large extent on the accuracy of the transistor models and the associated input parameters. Similarly, process and device engineers also have a strong incentive to familiarize themselves with transistor models, because transistor and process design are intimately interrelated with circuit design. For example, in digital circuits the switching speed of logic gates can only be maximized if the design of the process, transistor and circuits are each fully optimized. The relationship between these three important elements of the design procedure will be discussed in Chapter 7 for the important case of ECL circuits.

In this chapter we will consider the transistor models that are used in widely available computer-aided circuit design programs such as SPICE [1]. The simple DC Ebers–Moll model will be used as a starting point, and additional physical mechanisms added to the basic model as required. In this way, the full Gummel–Poon bipolar transistor model will evolve in a number of well-defined and easy to understand stages. Consideration will also be given to the important transistor parameters, especially the forward transit time τ_F, which defines the fundamental switching speed of a bipolar transistor. Important parasitics, such as base resistance and collector capacitance, will also be described, as these have a strong influence on bipolar circuit performance. Finally, our study of bipolar transistor models will conclude with a detailed description of the SPICE bipolar transistor model.

3.2 EBERS–MOLL MODEL

3.2.1 Basic DC Model

The Ebers–Moll model [2] is a simple, large-signal model for describing the DC behaviour of a bipolar transistor. The model configuration is illustrated in Figure 3.1 for an *npn* transistor. Equations for the forward and reverse diode currents I_F and I_R are needed to complete the model, and these are given by:

$$I_F = I_{ES} \left(\exp \frac{qV_{BE}}{KT} - 1 \right) \tag{3.1}$$

$$I_R = I_{CS} \left(\exp \frac{qV_{BC}}{KT} - 1 \right) \tag{3.2}$$

A third equation, termed the reciprocity relation, links the saturation currents I_{ES} and I_{CS} to the common base current gains [2]:

$$\alpha_F I_{ES} = \alpha_R I_{CS} = I_S \tag{3.3}$$

From these equations it is clear that three parameters are needed to fully describe a transistor in the Ebers–Moll model, namely α_F, α_R and I_S. A simple Gummel plot measurement, as described in Chapter 1, is adequate to provide experimental values for these parameters. The terminal currents of the transistor can be easily expressed in terms of the three transistor parameters:

$$I_C = \alpha_F I_F - I_R \tag{3.4}$$

$$I_E = \alpha_R I_R - I_F \tag{3.5}$$

$$I_B = (1 - \alpha_F) I_F + (1 - \alpha_R) I_R \tag{3.6}$$

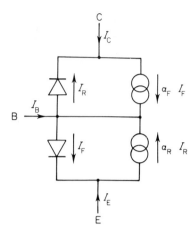

Figure 3.1. The basic DC Ebers–Moll model of a bipolar transistor

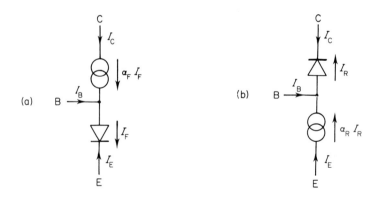

Figure 3.2. Simplifications of the basic DC Ebers–Moll model for transistors operating in (a) the forward active region and (b) the reverse active region

The Ebers–Moll model is firmly based on the physics of the device operation, as can be seen by referring to the theory in Chapter 2. The ideal diodes provide the expected exponential relationship between current and base/emitter voltage, and the current generators describe the transistor action that takes place because of the very narrow base region.

All the components in Figure 3.1 are required to model a transistor in saturation, but in the forward and reverse active regions considerable simplifications can be made. These are illustrated in Figure 3.2. Physically, in the forward active region the current I_F is the total current crossing the emitter/base junction (($I_{ne} + I_{pe}$) in Figure 2.1), while $\alpha_F I_F$ is the electron current at the edge of the collector/base depletion region (I_{nc}). Recombination in the emitter/base depletion region (I_{rg}) is not modelled in the basic Ebers–Moll model.

3.2.2 Non-linear Hybrid-Π Model

For the purposes of computer simulation a change in the form of the Ebers–Moll model is desirable. The reasons for this are not immediately apparent, but relate to the modelling of second-order mechanisms such as recombination in the emitter/base depletion region. A more suitable form of the Ebers–Moll model, the non-linear hybrid-Π model, is illustrated in Figure 3.3

In the non-linear hybrid-Π model the common component I_{CT} of the emitter and collector currents (equations (3.4) and (3.5)) has been identified and extracted:

$$I_{CT} = I_S \left[\left(\exp \frac{qV_{BE}}{KT} - 1 \right) - \left(\exp \frac{qV_{BC}}{KT} - 1 \right) \right] \quad (3.7)$$

Using equations (1.8) and (3.3), equations (3.4) and (3.5) can be rewritten as:

$$I_E = -I_{CT} - \frac{I_S}{\beta_F} \left(\exp \frac{qV_{BE}}{KT} - 1 \right) \quad (3.8)$$

$$I_C = I_{CT} - \frac{I_S}{\beta_R} \left(\exp \frac{qV_{BC}}{KT} - 1 \right) \quad (3.9)$$

It can be clearly seen that these equations for the emitter and collector currents are the same as those predicted by the model in Figure 3.3. This serves to emphasize that the non-linear hybrid-Π model is merely a rearrangement of the form of the Ebers–Moll model. Three parameters are needed to characterize a bipolar transistor in the non-linear hybrid-Π model, and these are β_F, β_R and I_S.

In the forward and reverse active regions the model can be considerably simplified, as illustrated in Figure 3.4. Physically, in the forward active region I_{CT} represents the electron current at the edge of the collector/base depletion

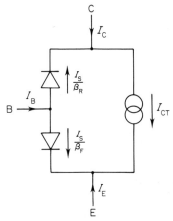

Figure 3.3 The non-linear hybrid-Π version of the Ebers–Moll model

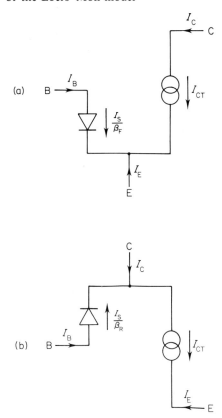

Figure 3.4 Simplifications of the non-linear hybrid-Π model for transistors operating in (a) the forward active region and (b) the reverse active region

region (I_{nc} in Figure 2.1), which of course is the component of current that traverses directly from the emitter to the collector.

3.2.3 AC Ebers–Moll Model

The bipolar transistor models discussed so far are only suitable for modelling the DC characteristics of bipolar transistors. In order to model AC effects, charge storage in the device must be described, and this requires the incorporation of additional parameters. These can be characterized into three broad types: series ohmic resistances, depletion capacitances and charge storage capacitances due to the mobile carriers in the transistor (diffusion capacitances). Figure 3.5 shows how the DC model can be extended to include these additional parameters. Note that internal nodes E′, B′ and C′ have been defined. In this extended model the emitter and collector diode currents are defined in terms of these internal node voltages.

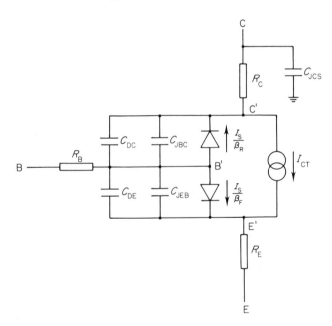

Figure 3.5. The basic AC Ebers–Moll model of a bipolar transistor

(a) Series resistances R_C, R_E and R_B

The resistances R_C, R_E and R_B represent the series resistance of the semiconductor between the active transistor area and the emitter, collector and base contacts. These are illustrated schematically in Figure 3.6 for a typical planar bipolar transistor.

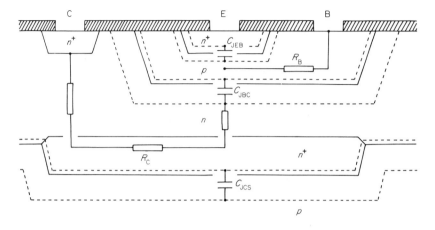

Figure 3.6. Cross-sectional view of a bipolar transistor showing the origins of the parasitic resistances and capacitances

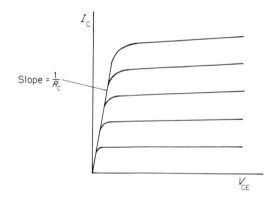

Figure 3.7. Measurement of the collector resistance R_C from the transistor output characteristics

The collector resistance R_C can be considered as the sum of the series resistance of the epitaxial layer below the active transistor area, the buried layer and the epitaxial layer below the collector contact. Its effect on the transistor characteristics is to decrease the slope of the initial part of the characteristic, as illustrated in Figure 3.7. This provides a simple way of measuring the collector resistance, though care must be taken because R_C is, in practice, a function of collector current and collector/base voltage [2]. The value shown in Figure 3.7 is for a transistor operating in strong saturation. A higher value is obtained when the transistor is operating in the forward active region [2].

Series resistance in the emitter is very small in most practical bipolar transistors because the emitter is very heavily doped. The main component of emitter resistance is generally contact resistance, which is of the order of

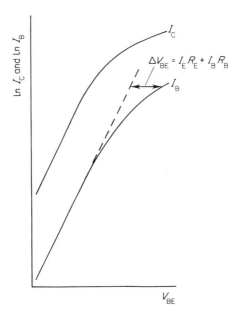

Figure 3.8. Gummel plot showing the effect of series emitter and base resistance

$5 \times 10^{-7}\,\Omega\text{cm}^2$. Its effect on the transistor is to reduce the voltage seen at the emitter/base junction by a factor $I_E R_E$, as illustrated in Figure 3.8. Although emitter resistance is small in most practical transistors, in polysilicon emitters it can be very large [3]. Emitter resistance will therefore be considered in more detail in the next chapter.

Base resistance is the most important of the series resistances, since it has a strong influence on the switching speed of the bipolar transistor. Its effect on the DC base and collector currents is similar to that of emitter resistance, as can be seen from Figure 3.8. Accurate specification and measurement of base resistance is very difficult, because it depends strongly on the operating conditions of the transistor [4]. This topic will be considered in more detail later in this chapter.

(b) Depletion capacitances C_{JEB}, C_{JBC} and C_{JCS}

The fixed charges in the depletion regions of the emitter/base and collector/base junctions give rise to capacitances, denoted by C_{JEB} and C_{JBC}, respectively. Figure 3.5 demonstrates how these can be incorporated into the transistor model and Figure 3.6 shows the physical origin of the capacitances in a planar transistor. Also shown is the substrate capacitance C_{JCS}.

The depletion capacitances are a non-linear function of voltage, and are modelled by physics-based [5] equations of the form:

$$C_{JEB} = C_{JE}\left[1 - \frac{V_{BE}}{V_{JE}}\right]^{-M_{JE}} \quad (3.10)$$

where C_{JE} is the value of the emitter/base capacitance at zero bias, V_{JE} is the junction built-in voltage and M_{JE} is a factor that defines the gradient of the emitter profile. For an abrupt pn junction $M_{JE} = 0.5$, and for a linear-graded junction $M_{JE} = 0.33$ [5]. In practical devices M_{JE} is likely to have a value between these two limits. The parameters C_{JE}, V_{JE} and M_{JE} can, in principle, be obtained by fitting equation (3.10) to the measured capacitance/voltage relationship, though in practice this is often difficult. A similar pair of equations is used to model C_{JBC} and C_{JCS}, thereby introducing a further six parameters: C_{JC}, V_{JC}, M_{JC}, C_{JS}, V_{JS} and M_{JS}.

(c) Diffusion capacitances C_{DE} and C_{DC}

The diffusion capacitances model the charge due to the mobile carriers in the transistor. This charge can be conveniently partitioned into two parts, one associated with the forward-biased emitter junction and one with the forward-biased collector junction.

The charge associated with the emitter can be calculated by forward biasing the emitter/base junction and putting zero bias across the collector/base junction. Figure 2.2(c) shows typical minority carrier distributions in the emitter and base for these bias conditions. Minority carriers are also present in the emitter/base and collector/base depletion regions. The total stored charge can therefore be written as the sum of these individual minority carrier charges:

$$Q_{DE} = Q_E + Q_{EBD} + Q_B + Q_{CBD} \quad (3.11)$$

where Q_E, Q_{EBD}, Q_B and Q_{CBD} represent the minority carrier charges in the emitter, emitter/base depletion region, base and collector/base depletion region, respectively. For modelling purposes the charge associated with the forward-biased emitter can therefore be written as:

$$Q_{DE} = (\tau_E + \tau_{EBD} + \tau_B + \tau_{CBD})I_{CC} \quad (3.12)$$

where τ_E is the emitter delay, τ_{EBD} the emitter/base depletion region transit time, τ_B the base transit time, and τ_{CBD} the collector/base depletion region transit time. These parameters will be discussed in more detail in Section 3.6. I_{CC} is the component of I_{CT} which is associated with the emitter/base junction:

$$I_{CC} = I_S\left(\exp\frac{qV_{BE}}{KT} - 1\right) \quad (3.13)$$

Finally, the delays in the individual regions of the device can be summed to give:

$$Q_{DE} = \tau_F I_{CC} \qquad (3.14)$$

where τ_F is the forward transit time.

A similar analysis can be carried out for the case where the collector/base junction is forward biased and the emitter/base junction zero biased. The minority carrier charge resulting from the forward-biased collector can be summed to give:

$$Q_{DC} = \tau_R I_{EC} \qquad (3.15)$$

where τ_R is the reverse transit time and I_{EC} is the component of I_{CT} that is associated with the collector/base junction:

$$I_{EC} = I_S \left(\exp \frac{qV_{BC}}{KT} - 1 \right) \qquad (3.16)$$

For a transistor in saturation both emitter and collector junctions are forward biased. The total minority carrier charge in the transistor can therefore be calculated by assuming that superposition applies. In other words, the total minority carrier charge is assumed to be equal to the sum of the charge due to each junction acting separately, i.e. $Q_{DE} + Q_{DC}$. These charges can then be related to the non-linear diffusion capacitances C_{DE} and C_{DC} by:

$$C_{DE} = \frac{Q_{DE}}{V_{B'E'}} \qquad (3.17)$$

$$C_{DC} = \frac{Q_{DC}}{V_{B'C'}} \qquad (3.18)$$

Two parameters are needed to model the diffusion capacitance, namely the forward and reverse transit times τ_F and τ_R.

3.3 SMALL-SIGNAL HYBRID-Π MODEL

The majority of the elements in Figure 3.5 are non-linear, and hence circuit analysis can only proceed with the aid of a computer. However, in circuits where the AC signal excursions around the DC quiescent operating point are small it is possible to approximate the non-linear elements by linear ones. This approximation is referred to as small-signal operation, and it applies to a variety of analogue circuits, amplifiers being one example. Digital circuits, which operate over the full range of operating conditions from cut-off to saturation, can only be modelled with the aid of the full non-linear model.

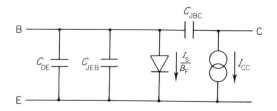

Figure 3.9. Simplified version of the AC Ebers–Moll model from which the small-signal hybrid-π model can be derived

For small-signal applications the model in Figure 3.5 can be simplified and linearized to give a small-signal equivalent circuit. In the forward active region the emitter/base junction is forward biased and the collector/base reverse biased. There is therefore no charge storage associated with the collector junction. This leads to the simplified model in Figure 3.9. The series resistances R_C, R_E and R_B and the substrate capacitance C_{JCS} have also been neglected in this simplified circuit.

The forward-biased emitter/base diode in Figure 3.9 can be linearized to an equivalent input resistance r_π, which can be calculated by differentiating the base current with respect to the base/emitter voltage:

$$I_B = \frac{I_S}{\beta_F} \exp \frac{qV_{BE}}{KT} \tag{3.19}$$

$$\frac{\partial I_B}{\partial V_{BE}} = \frac{I_S}{\beta_F} \cdot \frac{q}{KT} \cdot \exp \frac{qV_{BE}}{KT} = \frac{qI_B}{KT} \tag{3.20}$$

$$r_\pi = \frac{\partial V_{BE}}{\partial I_B} = \frac{KT}{qI_B} \tag{3.21}$$

Similarly, the current generator in Figure 3.9 can be linearized by differentiating the collector current with respect to the base/emitter voltage:

$$\frac{\partial I_C}{\partial V_{BE}} = g_m = I_S \frac{q}{KT} \exp \frac{qV_{BE}}{KT} = \frac{qI_C}{KT} \tag{3.22}$$

where g_m is the transconductance of the transistor.

Finally, the emitter diffusion capacitance C_{DE} can be linearized to:

$$C_{DE} \text{ (small signal)} = \frac{dQ_{DE}}{dV_{BE}} = \tau_F \frac{q}{KT} I_S \exp \frac{qV_{BE}}{KT} = g_m \tau_F \tag{3.23}$$

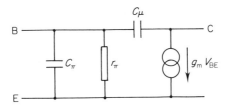

Figure 3.10. Small-signal hybrid-π model of a bipolar transistor

The resulting small-signal equivalent circuit is shown in Figure 3.10. The capacitor C_μ is the collector/base depletion capacitance and C_π is the sum of the emitter/base depletion capacitance and the emitter diffusion capacitance:

$$C_\pi = C_{JEB} + g_m \tau_F \tag{3.24}$$

This model is the well-known small-signal hybrid-π model [2].

3.4 GUMMEL–POON MODEL

The Gummel–Poon model [6] was introduced in 1970 and is an improved version of the AC Ebers–Moll model in Figure 3.5. Three second-order, high-level effects are modelled in an elegant and unified way. In this way, important parameters for the AC device behaviour also shape the DC characteristics of the transistor. The effects included are:

(1) High-level injection;
(2) Basewidth modulation;
(3) The variation of τ_F with I_C [7,8].

The Gummel–Poon model has been described in detail in the literature [2,6], and hence in this section we will merely state the relevant model equations, without considering their derivation. This will allow the emphasis of this section to be directed towards explanations of the underlying physical justification of the equations. The interested reader is referred to the book by Getreu [2] for more mathematical derivations of the equations.

The essence of the Gummel–Poon model is a new definition of the current I_{CT} (Figure 3.5) in terms of the internal physics of the transistor:

$$I_{CT} = \frac{I_S}{Q_B}\left[\left(\exp\frac{qV_{B'E'}}{KT} - 1\right) - \left(\exp\frac{qV_{B'C'}}{KT} - 1\right)\right] \tag{3.25}$$

where Q_B is the majority carrier charge in the base, normalized to the zero-bias majority carrier charge in the base. At zero bias Q_B is therefore equal to unity, and equation (3.25) reduces to the relevant Ebers–Moll equation (equation (3.7)). On application of bias to the junctions, Q_B takes on values other than unity. This provides a means of modelling basewidth modulation, high-level injection and the variation of τ_F with I_C, since all these mechanisms modulate the majority carrier charge in the base.

The normalized majority carrier charge in the base Q_B is given by an equation of the form:

$$Q_B = \frac{Q_1}{2} + \left[\left(\frac{Q_1}{2}\right)^2 + Q_2\right]^{1/2} \tag{3.26}$$

where

$$Q_1 = \left[1 - \frac{V_{B'C'}}{V_{AF}} - \frac{V_{B'E'}}{V_{AR}}\right]^{-1} \tag{3.27}$$

and

$$Q_2 = \frac{BI_S}{I_{KF}}\left(\exp\frac{qV_{B'E'}}{KT} - 1\right) + \frac{I_S}{I_{KR}}\left(\exp\frac{qV_{B'C'}}{KT} - 1\right) \tag{3.28}$$

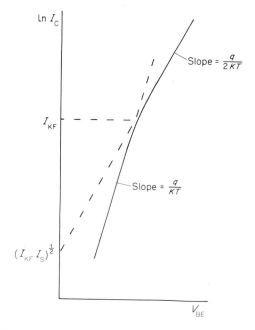

Figure 3.11. Gummel plot showing the measurement of the knee current I_{KF}, which defines the onset of high-level injection

V_{AF} is the forward Early voltage as defined in Figure 2.25 and V_{AR} is an equivalent reverse Early voltage, which needs to be modelled when the emitter/base junction is reverse biased. I_{KF} is the forward knee current which defines the onset of high-level injection, as illustrated in Figure 3.11. I_{KR} is an equivalent reverse knee current. The parameter B is used to model base-widening effects [7,8], which cause the effective basewidth of the transistor to increase at high current densities. This mechanism will be discussed in greater detail later in this chapter, and in the meantime it can be assumed that B is equal to unity.

The physical significance of equations (3.26)–(3.28) can be understood by considering the simplified case of a device in the forward active region. In this case equations (3.25), (3.27) and (3.28) reduce to:

$$I_{CT} = \frac{I_S}{Q_B}\left(\exp\frac{qV_{B'E'}}{KT} - 1\right) \tag{3.29}$$

$$Q_1 = \left[1 - \frac{V_{B'C'}}{V_{AF}}\right]^{-1} \tag{3.30}$$

$$Q_2 = \frac{I_S}{I_{KF}}\left(\exp\frac{qV_{B'E'}}{KT} - 1\right) \tag{3.31}$$

We will first consider the case of high-level injection. The criterion for the onset of high-level injection in the Gummel–Poon model is:

$$Q_2 \gg \frac{Q_1^2}{4} \tag{3.32}$$

Under high-level injection conditions the normalized majority carrier charge in the case Q_B can therefore be approximated by:

$$Q_B \simeq \sqrt{Q_2} \tag{3.33}$$

Substituting equations (3.33) and (3.31) into equation (3.29) yields:

$$I_{CT} = (I_S I_{KF})^{1/2} \exp\frac{qV_{BE}}{2KT} \tag{3.34}$$

This equation gives the expected $\exp(qV_{BE}/2KT)$ dependence of the collector current in the high-level injection regime, as predicted by equation (2.87) in Chapter 2. The intercept with the current axis is $(I_S I_{KF})^{1/2}$, as shown in Figure 3.11.

When the device is operating in low-level injection $Q_2 \ll Q_1^2/4$ and Q_B is approximately equal to Q_1. Using this approximation and substituting equation (3.30) into equation (3.29) gives:

$$I_{CT} = I_S \left(1 - \frac{V_{B'C'}}{V_{AF}}\right)\left(\exp\frac{qV_{B'E'}}{KT} - 1\right) \qquad (3.35)$$

This equation has the required $\exp(qV_{BE}/KT)$ dependence, but is multiplied by the term in parentheses which models basewidth modulation. The physical significance of this additional term is illustrated in Figure 3.12. The collector current at a given collector/emitter voltage is the sum of the collector current at zero collector/base volts $I_C(0)$ and that due to basewidth modulation I_{CBM}. From Figure 3.12 it can be seen that this is given by:

$$I_C = I_C(0) + I_{CBM} = I_C(0) + I_C(0)\frac{V_{CB}}{V_{AF}} \qquad (3.36)$$

$$= I_C(0)\left(1 + \frac{V_{CB}}{V_{AF}}\right)$$

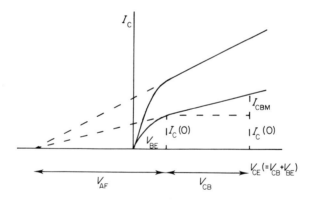

Figure 3.12. Transistor output characteristics illustrating the calculation of the forward Early voltage V_{AF}

Equation (3.36) is exactly the same form as the Gummel–Poon model equation (3.35). The change in sign comes about because in one case the collector/base voltage is defined with respect to the base, while in the other it is defined with respect to the collector.

The Gummel–Poon model requires five additional parameters, namely V_{AF}, V_{AR}, I_{KF}, I_{KR} and B. Also, in measuring the saturation current I_S the Gummel plot must be taken at a collector/base voltage of 0 V in order that basewidth modulation is properly modelled. The first four model parameters can easily be measured from transistor output characteristics and Gummel plots. The fifth parameter, B, will be considered later in this chapter.

3.5 MODELLING THE LOW-CURRENT GAIN

As discussed in Chapter 2, recombination of minority carriers in the emitter/base depletion region gives rise to a non-ideal, $\exp(qV_{BE}/mKT)$ dependence of the base current. In the models discussed so far this effect has been ignored. However, as described in Chapter 2, the recombination current can be regarded as being independent of the diffusion current [9], and hence can be simply modelled by adding an extra non-ideal diode to the Ebers–Moll or Gummel–Poon models. It should be emphasized that this is an empirical correction to the model equations, although, of course, the form of the correction is firmly rooted in the device physics.

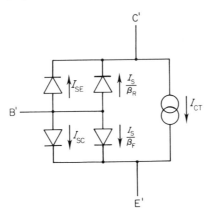

Figure 3.13. Modelling of the low-current gain in the non-linear hybrid-π version of the Ebers–Moll model

Figure 3.13 shows how recombination in the depletion region can be modelled by the inclusion of two non-ideal diodes. The first models recombination in the forward-biased emitter/base depletion region, and takes the form:

$$I_{gre} = I_{SE}\left(\frac{qV_{B'E'}}{N_E KT} - 1\right) \quad (3.37)$$

where N_E and I_{SE} are the emitter/base recombination ideality factor and saturation current, respectively. The second diode models recombination in the collector/base depletion region, and takes the form:

$$I_{grc} = I_{SC}\left(\exp\frac{qV_{B'C'}}{N_C KT} - 1\right) \quad (3.38)$$

where N_C and I_{SC} are the appropriate model parameters for the collector/base junction.

Four model parameters are needed to completely specify the low-current gain, and these are I_{SE}, N_E, I_{SC} and N_C.

3.6 FORWARD TRANSIT TIME τ_F

The forward transit time models the excess charge stored in the transistor when its emitter/base junction is forward biased and its collector/base junction zero biased. This is an extremely important parameter, since it provides a fundamental physical limit to the switching speed and maximum frequency of operation of a bipolar transistor. In this section we will therefore consider this parameter in more detail, beginning with a study of the components of τ_F and moving on to derive its relationship to the cut-off frequency f_T.

3.6.1 Components of τ_F

As discussed in Section 3.2, the forward transit time τ_F can be written as the sum of the individual delay times in the various regions of the transistor:

$$\tau_F = \tau_E + \tau_{EBD} + \tau_B + \tau_{CBD} \tag{3.39}$$

where τ_E and τ_{EBD} are associated with the excess minority carrier charge in the neutral emitter and emitter/base depletion regions, respectively. These two terms are generally small compared with the other terms in equation (3.39), although in high-speed bipolar transistors they can contribute significantly to the total forward transit time [10]. τ_B is associated with the excess minority carrier charge in the base and is frequently referred to as the base transit time. τ_{CBD} is the collector/base depletion layer delay, and in high-speed bipolar transistors is often of a similar magnitude to the base transit time.

The base transit time τ_B is given by [5,11]:

$$\tau_B = \frac{W_B^2}{\eta D_{nb}} \tag{3.40}$$

where $\eta = 2$ for a transistor with a uniformly doped base. For a non-uniformly doped base the variation of doping concentration in the base gives rise to a built-in electric field across the neutral base region. For example, in ion-implanted or diffused silicon bipolar transistors the base doping decreases on going from the emitter to the collector. This gives rise to an electric field that aids the transport of electrons across the base, with the result that the base transit time decreases. This can be taken into account in equation (3.40) by using a value of η greater than 2.

Equation (3.40) shows that the base transit time is proportional to the basewidth squared. In the design of high-speed bipolar transistors there is therefore a strong incentive to produce transistors with as small a basewidth as possible. In modern transistors the base transit time has decreased to such an extent that the collector/base depletion layer delay τ_{CBD} has become important. This delay is given by [12]:

$$\tau_{CBD} = \frac{W_{CBD}}{2v_{scl}} \tag{3.41}$$

where W_{CBD} is the width of the collector/base depletion region and v_{scl} is the scattering limited velocity. In silicon at 300 K V_{scl} has a value of 1×10^7 cm/s [13], while in gallium arsenide [13,14] it peaks at a value of 2×10^7 cm/s at an electric field of 3×10^3 V/cm, but decreases to a value of 6×10^6 V/cm at higher fields. From equation (3.41) it can be seen that the collector/base depletion layer delay can be reduced by decreasing the width of the collector/base depletion region. This can be achieved by increasing the doping concentration in the collector.

3.6.2 Base-widening Effects

The above analysis suggests that the forward transit time should be constant and independent of current. Although this is true at low currents, at high currents τ_F increases markedly with collector current [7,8]. The reason for this is an increase in the effective basewidth of the transistor [7] due to a current-dependent build-up of the minority carrier charge in the collector/base depletion region. This occurs when the mobile charge in the collector/base depletion region becomes greater than the fixed ionized charge, and this leads to the spreading of the neutral base region into the collector at high current densities. Two-dimensional spreading effects, as described by Van der Ziel and Agouridis [8], can also contribute to this degradation of τ_F at high collector currents.

The base-widening effect can be best understood by considering Poisson's equation (equation (2.13)) as applied to the collector/base depletion region:

$$\frac{dE}{dx} = \frac{\rho}{\varepsilon_0 \varepsilon_r} = \frac{q}{\varepsilon_0 \varepsilon_r} [p - n + N(x)] \qquad (3.42)$$

where p and n represent the mobile charge and $N(x)$ the fixed ionized charge in the depletion region. In an npn transistor the current is carried predominantly by electrons, and hence $p \simeq 0$ in equation (3.42). In the collector/base depletion region the electrons are transported by a drift mechanism, and this gives rise to an electron current density of:

$$J_n = qn\mu_n E = qnv_n \qquad (3.43)$$

where v_n is the electron drift velocity. At high electric fields such as are found in a reverse-biased collector/base junction the drift velocity saturates at the scattering limited velocity v_{scl}. Substituting equation (3.43) into equation (3.42) and making the above approximations yields:

$$\frac{dE}{dx} = \frac{1}{\varepsilon_0 \varepsilon_r} \left[qN(x) - \frac{J_n}{v_{scl}} \right] \qquad (3.44)$$

In the simple theory it is assumed that the mobile charge in the depletion region J_n/V_{scl} is much smaller than the fixed charge $q.N(x)$. However, it can be seen from equation (3.44) that this is only true if $J_n \ll q.N(x).v_{scl}$. For a typical collector doping concentration of 1×10^{16} cm^{-3} this is equivalent to an electron current density of 1.6×10^4 A/cm^2. Many practical bipolar transistors operate at collector currents of around 2×10^4 A/cm^2, which is precisely the regime where base widening is important. Optimization of the forward transit time therefore requires a collector doping that is high enough to ensure that base-widening effects are suppressed.

Modelling of base-widening effects is generally achieved through the use of an analytical expression for the parameter B. A variety of expressions have been used in circuit-modelling programs [2,6], and the interested reader is referred to the literature for further details. Additional model parameters are therefore required to define the variation of τ_F with collector current.

3.6.3 Relationship between τ_F and f_T

The cut-off frequency f_T of a bipolar transistor is defined as the frequency at which the extrapolated common emitter, short-circuit load, small-signal current gain drops to unity. This definition is illustrated schematically in Figure 3.14. Although f_T is not normally used as an input parameter to transistor models

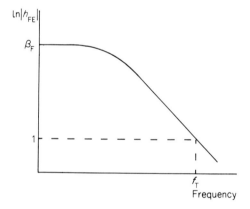

Figure 3.14. Variation of the small-signal common emitter current gain with frequency

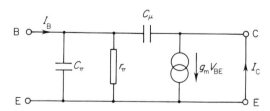

Figure 3.15. Use of the small-signal hybrid-π model for the calculation of the cut-off frequency f_T

it is frequently employed as a vehicle for measuring τ_F. It is also often quoted as a figure of merit for the AC performance of a bipolar transistor. In this section we will therefore derive the relationship between f_T and τ_F, and demonstrate how τ_F can be calculated from the measured value of f_T.

Since f_T is defined for small-signal conditions, the hybrid-π model in Figure 3.15 can be used to derive the common emitter current gain as a function of frequency. From this equivalent circuit the collector and base currents can be written as:

$$i_c = g_m V_{BE} - j\omega C_\mu V_{BE} \qquad (3.45)$$

$$i_b = V_{BE}(g_\pi + j\omega C_\pi \simeq j\omega C_\mu) \qquad (3.46)$$

where $g_\pi = 1/r_\pi$ and $\omega = 2\pi f$. The common emitter current gain can therefore be written as:

$$h_{FE} = \frac{i_c}{i_b} = \frac{g_m - j\omega C_\mu}{g_\pi + j\omega C_\pi + j\omega C_\mu} \qquad (3.47)$$

At most frequencies of practical interest, $g_m \gg j\omega C_\mu$, and hence equation (3.47) can be simplified to:

$$h_{FE} = \frac{\beta_F}{1 + j\omega r_\pi (C_\pi + C_\mu)} \qquad (3.48)$$

where we have used equations (3.21) and (3.22) to define β_F as:

$$\beta_F = g_m r_\pi \qquad (3.49)$$

It can be seen from equation (3.48) that the common emitter current gain approaches a value of β_F at low frequencies, as illustrated in Figure 3.14. At high frequencies the second term in the denominator of equation (3.48) is large with respect to unity, and β can be approximated by:

$$|h_{FE}| = \frac{\beta_F}{\omega r_\pi (C_\pi + C_\mu)} \qquad (3.50)$$

The common emitter current gain falls to unity when:

$$1 = \frac{\beta_F}{2\pi f_T r_\pi (C_\pi + C_\mu)} \qquad (3.51)$$

Rearranging and using equations (3.21) and (3.24) gives the following expression for the cut-off frequency:

$$2\pi f_T = \left[\tau_F + \frac{KT}{qI_C}(C_{JEB} + C_{JBC}) \right]^{-1} \qquad (3.52)$$

where τ_F is given by equation (3.39).

For completeness, an additional term should be included in equation (3.52) to account for the delay due to the series collector resistance and the collector/base capacitance [2]. The complete equation for the cut-off frequency then becomes:

$$2\pi f_T = \left[\tau_F + R_C C_{JBC} + \frac{KT}{qI_C}(C_{JEB} + C_{JBC}) \right]^{-1} \quad (3.53)$$

The collector delay term can be minimized by designing the transistor with a very thin epitaxial layer, so that the series resistance from the collector to the buried layer is minimized (Figure 3.6). An n^+ collector diffusion connecting the buried layer to the collector contact is also sometimes used to further decrease the collector resistance.

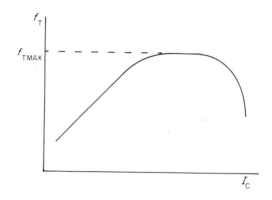

Figure 3.16. Variation of the cut-off frequency f_T with collector current

The dependence of f_T on collector current I_C is illustrated in Figure 3.16. At low currents the depletion capacitance term in equation (3.53) is much larger than the other two terms, and f_T increases with I_C. At medium currents the depletion capacitance term becomes smaller than τ_F, and hence f_T ceases to rise with collector current. In this part of the characteristic f_T is equal to f_{TMAX}, and τ_F is given by:

$$\tau_F = \frac{1}{2\pi f_{TMAX}} - R_C C_{JBC} \quad (3.54)$$

At high collector currents the cut-off frequency decreases markedly. This is caused by an increase in the value of τ_F due to the base-widening effects discussed in the previous section.

In many transistors a clearly defined region of constant-f_T is not discernable. However, in this case the forward transit time can be obtained from a graph

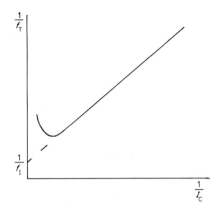

Figure 3.17. Method for measuring the maximum cut-off frequency f_{TMAX} of a bipolar transistor

Table 3.1. Breakdown of the components of f_T as a function of collector current density

Collector current density (A/cm^2)	Components of f_T (ps)					Cut-off frequency f_T (GHz)	Effective basewidth (μm)
	τ_E	τ_{EBD}	τ_B	τ_{CBD}	τ_{RE}		
2.9×10^2	0.23	2.2	0.53	3.0	45	3.1	0.043
1.1×10^3	0.25	1.4	0.59	3.0	14	8.3	0.043
3.6×10^3	0.32	1.0	0.86	3.0	5.9	14.4	0.043
1.3×10^4	0.43	0.70	1.40	3.0	3.1	18.4	0.049
2.4×10^4	0.58	0.53	1.80	4.4	2.7	15.9	0.057

of $1/f_T$ versus $1/I_C$, as illustrated in Figure 3.17. The intercept of the extrapolated straight line with the vertical axis can be used to calculate τ_F:

$$\tau_F = \frac{1}{2\pi f_I} - R_C C_{JBC} \tag{3.55}$$

In order to give some indication of the relative magnitudes of the terms in equation (3.53) Table 3.1 summarizes the components of f_T as a function of collector current. The figures shown are for a silicon high-speed bipolar transistor, and the results were computed using the BIPOLE device simulation program [15]. The delay τ_{RE} is the depletion capacitance term in equation (3.53), and the other delays are the components of τ_F.

At low collector currents it can be seen that τ_{RE} is by far the dominant component of f_T, as expected from equation (3.53). However, at collector currents around the peak f_T all the terms contribute significantly to the total delay, though τ_{RE}, τ_{CBD} and τ_B are the largest. The decrease in f_T at high collector currents is due to base-widening effects, as can be seen from the increase in the value of the effective basewidth.

Because of its importance in the measurment of τ_F, f_T is often used as figure of merit for the AC performance of a bipolar transistor. Unfortunately, it is not a realistic figure of merit for practical circuits, largely because it is measured with the output short-circuit. Also, it takes no account of base resistance and collector capacitance, which are very important in determining the transient behaviour of bipolar circuits. For this reason, an alternative figure of merit is often quoted, namely f_{MAX}. This is defined as the frequency at which the unilateral power gain falls to unity, and is given by [16]:

$$f_{MAX} = \left(\frac{f_T}{8\pi C_{JBC} R_B}\right) \quad (3.56)$$

It can be seen that this includes both base resistance and collector capacitance, and hence is a more realistic performance indicator. Other figures of merit have also been proposed [16], and the reader is referred to the literature for more information.

3.7 BASE RESISTANCE

Base resistance is one of the most important electrical parameters of a bipolar transistor. It limits the rate at which the input capacitance can be charged and is therefore one reason why bipolar transistors do not operate at the frequencies predicted by the values of forward transit time in Table 3.1. In many types of bipolar transistor it is useful to partition the base resistance into two parts, the intrinsic and extrinsic resistance, as illustrated in Figure 3.18. The total base resistance is then given by the sum of these two components.

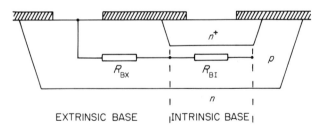

Figure 3.18. Cross-sectional view of a bipolar transistor illustrating the intrinsic and extrinsic components of the base resistance

The extrinsic base resistance R_{BX} is the resistance between the edge of the active transistor area and the base contact, and can easily be calculated from the transistor geometry and the extrinsic base sheet resistance R_{SBX}:

$$R_{BX} = \frac{R_{SBX}(b_b/l_b) + R_{CON}}{n_B} \quad (3.57)$$

where R_{CON} is the contact resistance, l_b and b_b are the geometrical length and width of the extrinsic base region and n_b is the number of base contacts.

The intrinsic base resistance R_{BI} is the resistance of the active base region, which is the region located beneath the emitter. It can be calculated from the transistor geometry and the intrinsic base sheet resistance R_{SBI} [17]:

$$R_{BI} = \frac{C R_{SBI}(b_e/l_e)}{n_B^2} \qquad (3.58)$$

where l_e and b_e are the geometrical emitter length and width and C is a constant which takes a value of 1/3 at low currents. It is immediately apparent from equations (3.57) and (3.58) that a large reduction of base resistance can be obtained if multiple-base contacts are used. However, this reduction is achieved at the cost of an increase in the transistor size and, more importantly, in the collector capacitance. In practice, therefore, an engineering compromise of two base contacts is generally chosen.

The intrinsic base sheet resistance is often referred to as the sheet resistance of the base under the emitter, or the resistance under the emitter, and can be measured using a pinch resistor. This is a type of resistor in which the emitter diffusion overlaps the base diffusion, as illustrated in Figure 3.19. The current through the resistor is therefore forced to flow through the high-resistance intrinsic base region underneath the emitter. A measure of the pinch resistance therefore gives a value for the intrinsic base resistance.

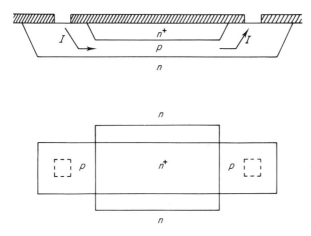

Figure 3.19. Measurement of the sheet resistance of the intrinsic base using a pinch resistor

The extrinsic base resistance is constant and very easy to model, but in many bipolar transistors the intrinsic base resistance is a strong function of current [18]. In particular, it decreases markedly at very high collector currents. Since this is precisely the regime in which many bipolar transistors are normally operated, it is important that this effect be properly understood and modelled.

The decrease in intrinsic base resistance at high collector currents is caused by current crowding due to the lateral flow of base current underneath the emitter, as illustrated in Figure 3.20. This results in an *IR* voltage drop along the width of the emitter, which causes the base/emitter voltage V_{BE} to vary with distance. The polarity of the voltage drop is such that V_{BE} is highest at the emitter edge closest to the base contact. Since the emitter current depends exponentially on V_{BE}, a small lateral voltage drop gives rise to a large variation in emitter current along the width of the emitter. In fact, the emitter current 'crowds' towards the base contact. In many practical bipolar transistors current crowding can cause the base resistance to decrease by a factor of 2 or more at high currents [18].

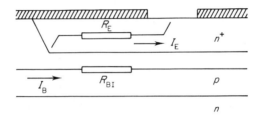

Figure 3.20. Illustration of current crowding in a bipolar transistor

Current crowding does not occur in all bipolar transistors, because it is strongly dependent on other factors, such as transistor geometry and emitter sheet resistance. For example, if the lateral voltage drop in the emitter, due to the sheet resistance of the emitter diffusion, is greater than that in the base, then current crowding will not occur. This situation is again shown in Figure 3.20, and illustrates that a lateral voltage drop in the emitter serves to counterbalance a similar drop in the base. Since the emitter current is approximately β times higher than the base current, the lateral voltage drop in the emitter can often be quite large.

Accurate modelling of current crowding requires a two-dimensional analysis. Although two-dimensional device simulation programs are available, they are very expensive in cpu time, and hence impractical for circuit simulation. Simple, empirical expressions for current crowding are therefore generally used in circuit simulation programs.

3.8 COLLECTOR/BASE CAPACITANCE

Collector capacitance is another parameter which is very important in determining the AC performance of a bipolar transistor. Its importance arises because of the Miller effect [19], in which the capacitance seen at the input of a bipolar transistor due to the collector/base capacitance is approximately

β times the collector/base capacitance. Clearly, in many practical bipolar transistors the Miller capacitance can be very large, and hence can seriously limit the performance of bipolar circuits. For this reason, high-speed bipolar transistors are carefully designed to have a very low collector capacitance.

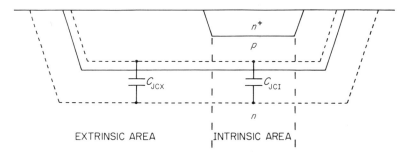

Figure 3.21. Cross-sectional view of a bipolar transistor illustrating the intrinsic and extrinsic components of the collector/base capacitance

For modelling purposes the collector capacitance is often partitioned into an intrinsic and extrinsic component, as illustrated in Figure 3.21. The intrinsic collector capacitance is determined primarily by the emitter geometry and the collector doping concentration. The emitter geometry is limited by lithography constraints, and the collector doping by the need to suppress high-current effects such as base widening. The transistor designer therefore has very little control over this component of the collector capacitance. The extrinsic collector capacitance is limited by the necessity of having space to make contact to the base of the transistor. It will be shown in Chapter 6 that self-aligned processing techniques [20] can be used to reduce considerably this component of the collector capacitance.

3.9 THE SPICE BIPOLAR TRANSISTOR MODEL

In this section we will complete our study of transistor modelling by considering a specific model of the bipolar transistor, namely that included in the SPICE program [1]. The SPICE circuit simulation program was introduced in 1973, and is now widely used throughout the world for the simulation of integrated circuits. It is basically a circuit simulator, which means that the device models need to be as simple as possible in order to minimize computational time, and hence allow relatively complex circuits to be modelled. The program has built-in device models for bipolar transistors, MOSFETS and JFETS, and input to these models is through sets of transistor parameters. The parameters for the SPICE bipolar transistor model are summarized in Table 3.2.

The SPICE bipolar transistor model is essentially a Gummel–Poon model, though base-widening effects are modelled slightly differently than originally proposed by Gummel and Poon [6]. Throughout this chapter the SPICE notation has been used in the model equations and in the labelling of the

Table 3.2. SPICE 2G bipolar transistor model parameters (after Nagel and Pederson [1])

Basic DC parameters

IS	Saturation current
BF	Maximum ideal forward gain
BR	Maximum ideal reverse gain
NF	Forward current ideality factor
NR	Reverse current ideality factor

Basic AC parameters

RC	Collector resistance
RE	Emitter resistance
RB	Low-current base resistance
IRB	Current where base resistance falls halfway to its maximum value
RBM	High-current base resistance
CJE	Emitter/base, zero bias depletion capacitance
VJE	Emitter/base built-in voltage
MJE	Emitter/base profile exponent
CJC	Base/collector, zero bias depletion capacitance
VJC	Base/collector built-in voltage
MJC	Base/collector profile exponent
XCJC	Fraction of B/C depletion capacitance connected to internal base node
CJS	Collector/substrate, zero bias capacitance
VJS	Collector/substrate built-in voltage
MJS	Collector/substrate profile exponent
FC	Coefficient for depletion capacitances in forward bias
TF	Forward transit time
TR	Reverse transit time

Gummel–Poon parameters

IKF	Knee current for roll-off of forward gain at high currents
IKR	Knee current for roll-off of reverse gain at high currents
VAF	Forward Early voltage
VAR	Reverse Early voltage
XTF	Coefficient for bias dependence of TF
VTF	Voltage describing VBC dependence of TF
ITF	Parameter for variation of TF at high currents
ISE	Saturation current for base/emitter leakage current
NE	Low-current forward current ideality factor
ISC	Saturation current for base/collector leakage current
NC	Low-current reverse current ideality factor

Additional parameters

EG	Semiconductor bandgap for temperature dependence of IS
XTI	Temperature exponent for effect on IS
XTB	Forward and reverse gain temperature exponent
PTF	Excess phase in g_m generator at frequency of $1/(2\pi \cdot \text{TF})$ Hz
KF	Flicker noise coefficient
AF	Flicker noise exponent

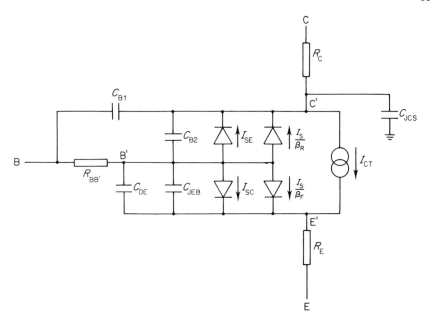

Figure 3.22. SPICE model of the bipolar transistor (after Nagel and Pederson [1])

model components. To a first approximation, therefore, the SPICE bipolar transistor model can be obtained by combining Figures 3.5 and 3.13, and using equations (3.25)–(3.40) to describe the non-linear behaviour of the components. The complete equivalent circuit is shown in Figure 3.22, and the collector and base currents are written as:

$$I_c = \frac{I_s}{Q_B}\left[\left(\exp\frac{qV_{B'E'}}{N_F KT} - 1\right) - \left(\exp\frac{qV_{B'C'}}{N_R KT} - 1\right)\right] - \frac{I_s}{\beta_R}\left(\exp\frac{qV_{B'C'}}{N_R KT} - 1\right)$$

$$- I_{SC}\left(\exp\frac{qV_{B'C'}}{N_C KT} - 1\right) \tag{3.59}$$

$$I_B = \frac{I_s}{\beta_F}\left(\exp\frac{qV_{B'E'}}{N_F KT} - 1\right) + \frac{I_s}{\beta_R}\left(\exp\frac{qV_{B'C'}}{N_R KT} - 1\right) + I_{SE}\left(\exp\frac{qV_{B'E'}}{N_E KT} - 1\right)$$

$$+ I_{SC}\left(\exp\frac{qV_{B'C'}}{N_C KT} - 1\right) \tag{3.60}$$

where Q_B is defined by equations (3.26)–(3.28). Two additional parameters N_F and N_R have been introduced to allow the exponents of the ideal emitter/base and collector/base diodes to be altered. In most practical bipolar transistors these would be set equal to unity. Also the parameter B in equation (3.28) is equal to unity, since SPICE does not model base widening using this parameter.

The SPICE base resistance model includes current crowding, and is given by:

$$R_{BB'} = R_{BM} + \frac{R_B - R_{BM}}{Q_B} \qquad (3.61)$$

At high currents Q_B is very large, and equation (3.61) reduces to $R_{BB'} = R_{BM}$. The parameter R_{BM} therefore represents the high-current value of base resistance, or the extrinsic base resistance. At low currents, Q_B is equal to Q_1, and in the absence of basewidth modulation Q_1 is equal to unity. Equation (3.61) therefore reduces to $R_{BB'} = R_B$. The parameter R_B is therefore the low-current value of base resistance, namely, the sum of the intrinsic and extrinsic base resistances. An alternative, more complicated, empirical expression for the base resistance is also available, which uses a parameter I_{RB}. When I_{RB} is not specified, equation (3.61) is used to model the base resistance.

The emitter/base depletion capacitance C_{JEB} is modelled using:

$$C_{JEB} = C_{JE} \left[1 - \frac{V_{B'E'}}{V_{JE}} \right]^{-M_{JE}} \qquad (3.62)$$

and similar expressions are used for the base/collector and collector/substrate depletion capacitances.

The emitter diffusion capacitance C_{DE} is modelled by:

$$C_{DE} = \frac{\partial}{\partial V_{B'E'}} \left[\frac{T_{FF} I_S}{Q_B} \left(\exp \frac{qV_{B'E'}}{N_F KT} - 1 \right) \right] \qquad (3.63)$$

where

$$T_{FF} = T_F \left[1 + X_{TF} \left(\frac{I_F}{I_F + I_{TF}} \right)^2 \exp \frac{V_{B'C'}}{1.44 \, V_{TF}} \right] \qquad (3.64)$$

and

$$I_F = I_S \left(\exp \frac{qV_{B'E'}}{N_F KT} - 1 \right) \qquad (3.65)$$

Equation (3.64) is an empirical expression for modelling the increase in τ_F at high currents, and it requires three model parameters: X_{TF}, I_{TF} and V_{TF}. This expression replaces the parameter B in the Gummel–Poon model described in Section 3.4. At low currents equation (3.64) reduces to $T_{FF} = T_F$, and hence T_F is the low-current forward transit time.

The base/collector depletion and diffusion capacitances have been combined in the SPICE model into a total base/collector capacitance C_{BC} given by:

$$C_{BC} = C_{JC}\left[1 - \frac{V_{B'C'}}{V_{JC}}\right]^{-M_{JC}} + T_R\left[\frac{qI_s}{N_R KT}\exp\frac{qV_{B'C'}}{KT}\right] \quad (3.66)$$

where

$$C_{B1} = C_{BC}(1 - X_{CJC}) \quad (3.67)$$

and

$$C_{B2} = C_{BC} X_{CJC} \quad (3.68)$$

The parameter X_{CJC} allows the distributed nature of the base resistance and base/collector capacitance to be modelled. When X_{CJC} is set equal to unity, C_{B2} becomes equal to the total base/collector capacitance C_{BC}. In this case, the modelling of the base/collector capacitance in the SPICE model of Figure 3.22 is equivalent to that in Figure 3.5. Alternatively, the collector capacitance can be partitioned into an extrinsic and intrinsic component. In this case, X_{CJC} can be chosen to make C_{B2} equal to the intrinsic collector capacitance and C_{B1} equal to the extrinsic collector capacitance.

In forward bias the depletion capacitances increase rapidly with forward voltage [2]. In fact, equation (3.10) gives an infinite capacitance when $V_{BE} = V_{JE}$. In order to avoid this computational hazard the emitter/base depletion capacitance reverts to the following form when $V_{BE} > F_C V_{JE}$:

$$C_{JEB} = \frac{C_{JE}}{(1 - F_C)^{M_{JE}}}\left[1 + \frac{M_{JE}(V_{BE} - F_C V_{JE})}{V_{JE}(1 - F_C)}\right] \quad (3.69)$$

F_C is a parameter which takes a value between 0 and 1.0. Analogous equations are also used for the base/collector and collector/substrate capacitances in forward bias.

A complete set of the 40 SPICE parameters is given in Table 3.2. The list has been partitioned into four parts: those concerned with the basic DC Ebers–Moll model (Section 3.2.1), those concerned with the AC Ebers–Moll model (Section 3.2.3), the additional parameters concerned with the Gummel–Poon model (Section 3.4) and a number of parameters that have not been considered in the above discussion. The interested reader is referred to the SPICE user manual for further information on these additional parameters.

References

1. L. W. Nagel and D. O. Pederson, 'Simulation program with integrated circuit emphasis', *16th Midwest Symposium on Circuit Theory*, 12 April (1973).
2. I. E. Getreu, *Modelling the Bipolar Transistor*, Elsevier, Amsterdam (1978).
3. E. F. Chor, P. Ashburn and A. Brunnschweiler, 'Emitter resistance of arsenic- and phosphorus-doped polysilicon emitter transistors', *IEEE Electron. Device Lett.*, **EDL6**, 516 (1985).

4. W. M. C. Sansen and R. G. Meyer, 'Characterization and measurement of the base and emitter resistances of bipolar transistors', *IEEE Jnl Solid State Circuits*, **SC7**, 492 (1972).
5. S. M. Sze, *Physics of Semiconductor Devices*, John Wiley, New York (1981).
6. H. K. Gummel and H. C. Poon, 'An integral charge control model of bipolar transistors', *Bell Syst. Tech. Jnl*, **49**, 827 (1970).
7. C. T. Kirk, 'A theory of transistor cut-off frequency falloff at high current densities', *IRE Trans. Electron. Devices*, **ED9**, 164 (1962).
8. A van der Ziel and D. Agouridis, 'The cut-off frequency fall-off in UHF transistors at high currents', *Proc IEEE*, **54**, 411 (1966).
9. P. Ashburn, D. V. Morgan and M. J. Howes, 'A theoretical and experimental study of recombination in silicon *pn* junctions', *Solid State Electronics*, **18** 569 (1975).
10. J. A. Kerr and F. Berz, 'The effect of emitter doping gradient in f_T in microwave bipolar transistors', *IEEE Trans. Electron. Devices*, **ED22**, 15 (1975).
11. J. Lindmayer and C. Wrigley, 'The high-injection level operation of drift transistors', *Solid State Electronics*, **2**, 79 (1961).
12. R. G. Meyer and R. S. Muller, 'Charge control analysis of the collector base space-charge region contribution to bipolar transistor time constant', *IEEE Trans. Electron. Devices*, **ED34**, 450 (1987).
13. P. Smith, M. Inoue and J. Frey, 'Electron velocity in Si and GaAs at very high electric fields', *App. Phys. Lett.*, **37**, 797 (1980).
14. J. G. Ruch and G. S. Kino, 'Measurement of the velocity field characteristics of gallium arsenide', *App. Phys. Lett.*, **10**, 40 (1967).
15. D. J. Roulston, S. G. Chamberlain and J. Sehgal, 'Simplified computer aided analysis of double diffused transistors including two-dimensional high-level effects', *IEEE Trans. Electron. Devices*, **ED19**, 809 (1972).
16. G. W. Taylor and J. G. Simmons, 'Figure of merit for integrated bipolar transistors', *Solid State Electronics*, **29**, 941 (1986).
17. R. M. Burger and R. P. Donovan, *Fundamentals of Silicon Integrated Device Technology*, Vol. II, Prentice-Hall, Englewood Cliffs, NJ (1968), p. 115.
18. J. E. Lary and R. L. Anderson, 'Effective base resistance of bipolar transistors', *IEEE Trans. Electron. Devices*, **ED32**, 2503 (1985).
19. P. R. Gray and R. G. Meyer, *Analog Integrated Circuits*, John Wiley, New York (1977), p. 383.
20. T. H. Ning, R. D. Isaac, P. M. Solomon, D. D. Tang, H. Yu, G. C. Feth and S. K. Wiedmann, 'Self-aligned bipolar transistors for high-performance and low power delay VLSI', *IEEE Trans. Electron. Devices*, **ED28**, 1010 (1981).

Chapter 4
POLYSILICON EMITTERS

4.1 INTRODUCTION

Polycrystalline silicon, or polysilicon, is a form of silicon part-way between perfectly ordered single-crystal silicon and totally unordered amorphous silicon. It consists of small, randomly oriented grains of single-crystal silicon, separated by disordered regions known as grain boundaries. This is illustrated schematically in Figure 4.1. Polysilicon is widely used in MOS processes to form the gate

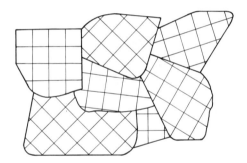

Figure 4.1. Schematic illustration of the structure of polycrystalline silicon

electrode of the MOS transistor. It is particularly useful in this type of application because of its low deposition temperature (600°C) and its ability to withstand the high temperatures routinely used in the fabrication of integrated circuits (900–1200°C). It also has useful electrical properties, since in many respects it behaves in a way similar to single-crystal silicon. Thus it can be doped to produce n- or p-type layers, and at high doping concentrations reasonably low sheet resistances can be achieved ($\simeq 20\,\Omega/\text{sq}$). This property makes it possible to use polysilicon as an interconnection layer in MOS integrated circuits.

Polysilicon was first used for the emitter of a bipolar transistor by Graul et al. [1] in 1976, and the fabrication sequence is illustrated in Figure 4.2. An undoped polysilicon layer is first deposited onto the wafer after the definition of an emitter window. Arsenic is then implanted into the polysilicon and

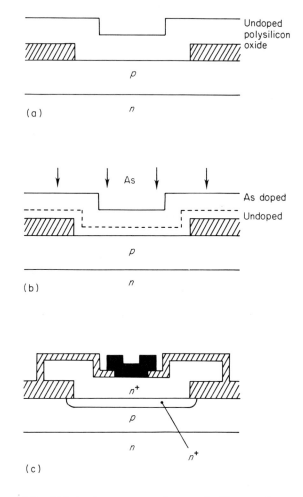

Figure 4.2. Fabrication sequence for a polysilicon emitter

subsequently diffused into the underlying single-crystal silicon at a temperature of around 900°C. Penetration of the emitter dopant into the single-crystal silicon is necessary in order to produce a good-quality device. However, the penetration can be extremely small; for example, good transistor characteristics have been reported in the literature [2] for emitter/base junction depths as shallow as 0.01 μm. The device is completed by defining the polysilicon and adding contact windows and metallization, as shown in Figure 4.2(c).

The fabrication sequence of Figure 4.2 is the most commonly used approach for producing polysilicon emitters. However, an alternative method has also been used and this is outlined in figure 4.3. In this scheme the emitter is formed in the traditional way by implanting arsenic into single-crystal silicon to produce an emitter/base junction approximately 0.2 μm deep. An arsenic-doped polysilicon layer is then deposited and defined prior to the emitter heat cycle

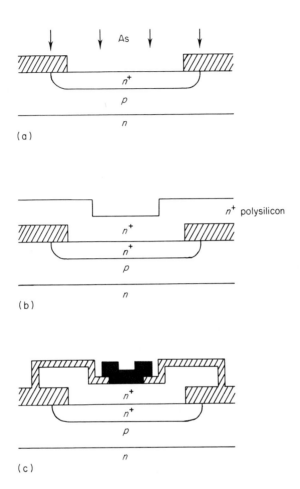

Figure 4.3. Fabrication sequence for a polysilicon contacted emitter

at a temperature of around 900°C. As before, the device is completed by adding contacts and metallization. This fabrication procedure produces a device which is essentially a conventional transistor with a polysilicon capping layer, and is often referred to as a polysilicon-contacted emitter.

The major advantage of polysilicon emitters is their suitability for producing shallow emitter/base junctions, and also their compatibility with self-aligned fabrication techniques. These techniques will be described in detail in Chapter 6, but in essence they allow the parasitic resistances and capacitances of a bipolar transistor to be minimized, with the result that a considerable improvement in circuit performance is obtained. A further important property of polysilicon emitters is that significantly higher values of common emitter current gains are obtained. For the fabrication sequence of Figure 4.2 gain improvements of up to a factor of 10 have been reported [4,5], whereas for the sequence of Figure 4.3 improvements by a factor of 2 or 3 are typical.

Table 4.1. Comparison of plane and peripheral components of emitter/base capacitance

Emitter size (μm)	Emitter/base junction depth (μm)	Emitter/base capacitance	
		Plane	Peripheral
1.5 × 1.5	0.2	4.1	3.0
	0.1	4.1	1.6
	0.02	4.1	0.4
0.5 × 0.5	0.2	0.46	0.99
	0.1	0.46	0.55
	0.02	0.46	0.14

The improved gain of the polysilicon emitter transistor is useful for a number of important reasons. First, the additional gain can be traded for an increase in the base doping concentration and hence a decrease in the base resistance [2]. As discussed in Chapter 3, this is likely to lead directly to an improvement in the switching speed of the bipolar transistor. More importantly, effective scaling of bipolar transistors to finer geometries requires a co-ordinated reduction in both the lateral and vertical dimensions [6]. The necessity for this can be understood by referring to the calculated values of emitter/base junction capacitance in Table 4.1. Here, the capacitance has been partitioned into a peripheral component due to the junction sidewalls and a plane component due to the bottom junction. For the purposes of this comparison the peripheral capacitances have been calculated assuming a cylindrical shape for the lateral diffusion. At an emitter geometry of 1.5 × 1.5 μm and a junction depth of 0.2 μm the peripheral capacitance contributes nearly 50%, of the total capacitance, and at a geometry of 0.5 × 0.5 μm this increases to nearly 70%. Effective scaling to smaller lateral geometries can therefore only be achieved if the junction depths are also scaled. As discussed in Section 2.6, reducing the emitter/base junction depth of a conventional bipolar transistor leads to an equivalent reduction in the current gain. This is in contrast to the situation for a polysilicon emitter, where the emitter/base junction depth can be reduced without any degradation of the current gain.

4.2 BASIC PHYSICS OF THE POLYSILICON EMITTER

The polysilicon emitter is an extremely complicated structure which is difficult to characterize experimentally and equally difficult to model. Two key elements of the emitter structure are responsible for these problems, namely, the polysilicon/silicon interface and the polysilicon grain boundaries. These two elements are crucial to the operation of the transistor, but unfortunately they are neither well characterized nor fully understood. In this section we will consider the physical mechanisms that occur in a polysilicon emitter, together with the experimental evidence for their importance. In the following section

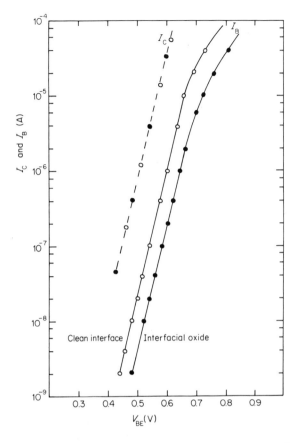

Figure 4.4. Gummel plots for practical polysilicon emitter transistors with and without a deliberately grown interfacial oxide (after Ashburn and Soerowirdjo [5], copyright ©1984 IEEE)

these mechanisms will be combined into a unified theory, while in Section 4.5 the practical aspects of polysilicon emitter design will be considered.

Experimental measurements on polysilicon emitter bipolar transistors have shown that the current gain is strongly influenced by the nature of the polysilicon/silicon interface [5]. Devices with a nominally clean interface, produced by a short etch in hydrofluoric acid prior to polysilicon deposition, have gains enhanced by a factor of 2 or 3. Alternatively, if a thin interfacial oxide is grown using a wet chemical treatment [4,5,7] gain enhancements of a factor of 10 or more are obtained [5]. Gummel plots for these two types of transistor are shown in Figure 4.4, where it can be seen that the improvement in gain results from a decrease in base current rather than an increase in collector current. The explanation for these results is that the presence of the interfacial layer suppresses hole injection from the base into the emitter. In other words,

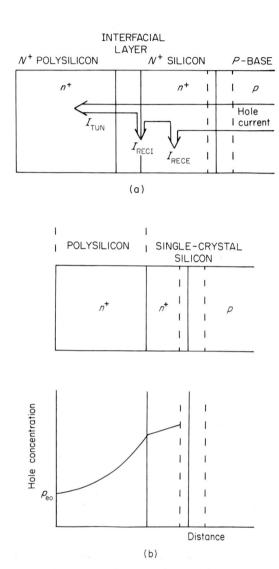

Figure 4.5. (a) Schematic illustration of the components of the base current of a polysilicon emitter transistor with an interfacial layer; (b) schematic illustration of the hole distribution in a polysilicon emitter transistor without an interfacial layer

current component I_{pe} in Figure 2.1 is significantly reduced. This happens because the holes must penetrate through the interfacial oxide in order to reach the emitter contact; the dominant mechanism for this transport through the oxide has been identified as tunnelling [4,5], as illustrated schematically in Figure 4.5(a). The hole current, and hence the base current, is therefore determined by the tunnelling properties of the interfacial oxide, in particular by the interfacial layer thickness and the effective barrier height for holes χ_h.

The above reasoning implies that the current gain of a polysilicon emitter transistor can be increased indefinitely by increasing the thickness of the interfacial layer. Clearly, this is unreasonable, and in practice the magnitude of the gain improvement is limited by other mechanisms. For example, some of the holes recombine in the single-crystal emitter, giving rise to an emitter recombination current I_{RECE} in Figure 4.5(a). In the limit when the single-crystal emitter is very thick no gain improvement is obtained, since all the holes recombine before reaching the interface. The single-crystal emitter depth must therefore be small with respect to the hole diffusion length in the emitter in order to obtain an improved gain. Holes can also recombine at surface states at the polysilicon/ silicon interface to give the current I_{RECI} in Figure 4.5(a). This component of base current is more difficult to quantify because of uncertainties in the nature of the interfacial layer. Recombination in the emitter/base depletion region and in the base [8] can also contribute to the base current, as described in Section 2.2.

In devices with a nominally clean interface improved gains are again obtained [3,5], although in this case the improvement is only by a factor of 2 or 3. These results can be explained by noting that in a polysilicon emitter transistor the effective emitter/base junction depth is given by the sum of the polysilicon and single-crystal emitter thicknesses. The transport properties of both the polysilicon and the single-crystal silicon regions therefore influence the base current. In a conventional, shallow-emitter transistor a linear hole distribution is obtained in the emitter (Figure 2.2(c)) because the junction depth is small with respect to the hole diffusion length. In a polysilicon emitter transistor the holes are not forced to recombine at the surface of the single-crystal silicon but are able to pass into the polysilicon layer and recombine there. This gives rise to a much shallower concentration gradient, as shown in Figure 4.5(b). Since the hole current I_{pe} is proportional to this gradient (equation (2.29)) it is evident that I_{pe} will be significantly smaller in a polysilicon emitter transistor.

The physics of polysilicon emitters is further complicated by the presence of the grain boundaries in the polysilicon and also by the large pseudo-grain boundary which is formed at the interface between the polysilicon and single-crystal silicon. These grain boundaries contain a high density of defects and dangling bonds, and hence can act as recombination centres for the minority carrier holes. In certain circumstances the grain boundaries can also block the transport of holes and hence enhance the current gain [3]. This occurs because the hole mobility in the vicinity of a grain boundary is considerably lower than that in single-crystal silicon [3]. This grain boundary blocking mechanism is particularly important in polysilicon emitters with a nominally clean interface.

The influence of grain boundaries on the electrical characteristics of polysilicon emitter transistors has been experimentally demonstrated and arsenic segregation identified as being important [9,10]. It has been shown by a number of researchers that arsenic segregates to polysilicon grain boundaries, and also that the segregated arsenic has a significant effect on the electrical properties of the grain boundary [11]. Patton *et al.* [10] have used these results to show that the base current of a polysilicon emitter transistor with a nominally clean interface decreases by approximately 40% when additional arsenic is segregated to the grain boundaries. Furthermore, Neugroschel *et al.* [9] have demonstrated that segregation at the pseudo-grain boundary at the polysilicon/silicon interface is particularly important in determining the current gain. The most probable explanation for these results is that the arsenic segregation is decreasing the density of trapping states at the polysilicon/silicon interface and hence reducing the amount of recombination. Alternatively, it has been proposed that the presence of the segregated arsenic at the interface could give rise to a low–high–low junction [12]. This would present a potential barrier to the transport of minority carrier holes across the interface and hence would suppress the base current. This explanation requires that the segregated dopant is electrically active, an assumption for which there is as yet no experimental evidence.

The above discussion has shown that both tunnelling and transport mechanisms contribute to the improved gains of polysilicon emitter transistors. Tunnelling offers by far the biggest gain improvement, but a deliberately grown interfacial layer is needed which can give rise to an undesirable increase in the emitter resistance of the transistor [13]. This, in turn, leads to a degradation of circuit performance [14], particularly at sub-micron geometries. The transport mechanism offers much lower gain improvements, but the benefit can be maximized if careful attention is paid to the mechanisms which limit the extent of the gain improvement. These are primarily recombination at the polysilicon/silicon interface, at grain boundaries and in the single-crystal emitter. The task of the transistor designer is therefore to engineer the require gain by carefully controlling the interfacial layer thickness, the emitter/base junction depth and the emitter doping profile.

4.3 THEORY OF POLYSILICON EMITTERS

4.3.1 Effective Recombination Velocity

The simplest method of modelling a polysilicon emitter is to use an approach analogous to that used for shallow conventional emitters, as given in equation (2.69). In this equation recombination at the metal contact was defined by the surface recombination velocity S_M and recombination in the emitter by the term in square brackets. It is therefore a simple matter to model a polysilicon emitter by replacing the surface recombination velocity S_M with an effective recombination velocity at the polysilicon/silicon interface S_{PI} [15], as illustrated in Figure 4.6. Equation (2.69) then becomes:

$$I_B = \frac{qAn_{io}^2}{G_e(W_E)+(N_{deff}(W_E)/S_{PI})}\left[1+\int_0^{W_E}\frac{[G_e(W_E)-G_e(x)]}{\tau_{pe}(x)N_{deff}(x)}dx + \frac{N_{deff}(W_E)}{S_{PI}}\int_0^{W_E}\frac{dx}{\tau_{pe}(x)N_{deff}(x)}\right]\exp\frac{qV_{BE}}{KT} \quad (4.1)$$

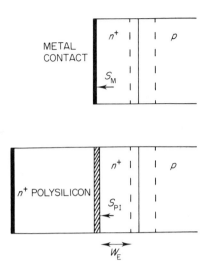

Figure 4.6. Comparison of a shallow emitter transistor and a polysilicon emitter transistor

S_{PI} includes transport and tunnelling mechanisms and recombination at the polysilicon/silicon interface. Recombination in the single-crystal emitter is modelled by the term in square brackets.

For practical polysilicon emitter transistors S_{PI} can be experimentally determined from the measured base current and the emitter doping profile by means of equation (4.1) [10]. This empirical approach has the advantage of simplicity, and allows the polysilicon emitter to be characterized using a single, experimentally measured parameter. Alternatively for an idealized polysilicon emitter transistor it is possible to derive a theoretical expression for S_{PI}, as will be shown in the next section. The unified approach of Eltoukhy and Roulston [16] and Yu *et al.* [15] will be used, so both tunnelling and transport mechanisms will be included. The modelling of the polysilicon layer will be covered first, followed by the interfacial layer and finally the single-crystal emitter.

4.3.2 Modelling the Polysilicon Layer

Several authors have proposed models for majority carrier conduction in polycrystalline silicon [17,18]. The grain boundaries are generally modelled as potential barriers, which result partly from the trapping of majority carriers at the dangling bonds and trapping states and partly from the disordered, amorphous nature of the grain boundary [17]. The reduced majority carrier mobility in polysilicon is then explained by the action of the potential barrier and the associated depletion region adjacent to the grain boundary.

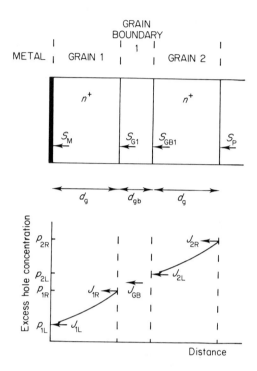

Figure 4.7. Schematic illustration of a polysilicon emitter showing the components of hole current in the polysilicon

In polysilicon emitters it is the minority carrier mobility that is of interest, and the extended state mobility model of Kim et al. [19] is more appropriate [15]. In this model the grain boundaries are treated as an amorphous layer of finite thickness, with a mobility that is considerably lower than that of the bulk silicon within the grain. The overall reduction in mobility is then explained by the lower mobility at the grain boundaries.

Figure 4.7 shows the assumed grain structure in our idealized polysilicon emitter. An effective recombination velocity is defined at each interface and these are given by equations of the form:

$$J_{1L} = qS_M P_{1L} \qquad (4.2)$$

where J_{1L} is the hole current density at the left side of the first grain and p_{1L} the excess hole concentration. This equation is identical in form to equation (2.47). The hole current density at either end of the first polysilicon grain can be calculated by solving the continuity equation for holes. This task is considerably simplified by noting that, in heavily doped polysilicon, the potential barrier due to carrier trapping is very small. The electric field inside the grain can therefore be neglected, and the approach of Section 2.3 used to solve for the hole current densities [15]:

$$J_{1L} = q(b_g p_{1R} - a_g p_{1L}) \quad (4.3)$$

$$J_{1R} = q(a_g p_{1R} - b_g p_{1L}) \quad (4.4)$$

where

$$a_g = \frac{D_{pp}}{L_{pp}} \coth \frac{d_g}{L_{pp}} \quad (4.5)$$

and

$$b_g = \frac{D_{pp}}{L_{pp}} \operatorname{csch} \frac{d_g}{L_{pp}} \quad (4.6)$$

D_{pp} and L_{pp} are the hole diffusion coefficient and diffusion length within the polysilicon grains and d_g is the width of the grain. Equating equations (4.2) and (4.3) gives the following relationship between the excess hole concentrations at the left and right edges of the first grain:

$$p_{1L} = \frac{b_g}{a_g + S_M} p_{1R} \quad (4.7)$$

Substituting into equation (4.4) then gives an expression for the hole current at the right edge of the first grain:

$$J_{1R} = q \left(a_g - \frac{b_g^2}{a_g + S_M} \right) p_{1R} \quad (4.8)$$

The effective recombination velocity at the right edge of the first grain S_{G1} is therefore given by:

$$S_{G1} = a_g - \frac{b_g^2}{a_g + S_M} \quad (4.9)$$

Recombination at dangling bonds and trapping states at the grain boundary is assumed to occur only at the interfaces between the grain boundary and the grains. This is equivalent to assuming that the grain is infinitesimally thin for the purposes of calculating the recombination but of finite thickness for calculating the effect of the grain boundary mobility. Recombination at the grain boundary is defined by a recombination velocity S_{GB}, which can be approximated by:

$$S_{GB} = \frac{N_{st}}{2} c_p v_{th} \qquad (4.10)$$

where N_{st} is the density of grain boundary traps, c_p the capture cross-section of holes and v_{th} the thermal velocity. It has been assumed that half of the traps are located at the left edge of the grain boundary and the other half at the right edge.

The grain boundary recombination velocity S_{GB} is defined as:

$$J_{GB} - J_{1R} = qS_{GB}p_{1R} \qquad (4.11)$$

$$J_{2L} - J_{GB} = qS_{GB}p_{2L} \qquad (4.12)$$

The current density J_{GB} at the left and right edges of the grain boundary can be calculated from the diffusion equation and Figure 4.7, and is given by:

$$J_{GB} = qD_{gb} \frac{(p_{2L} - p_{1R})}{d_{gb}} \qquad (4.13)$$

Inserting equation (4.13) into equations (4.11) and (4.12) gives the following equations for the hole currents at the two edges of the grain boundary:

$$J_{2L} = q(a_{gb}p_{2L} - b_{gb}p_{1R}) \qquad (4.14)$$

$$J_{1R} = q(b_{gb}p_{2L} - a_{gb}p_{1R}) \qquad (4.15)$$

where

$$a_{gb} = S_{GB} + \frac{D_{gb}}{d_{gb}} \qquad (4.16)$$

and

$$b_{gb} = \frac{D_{gb}}{d_{gb}} \qquad (4.17)$$

These linear equations are of the same form as equations (4.3)–(4.6), and this makes possible the simple extension of the effective recombination velocity concept, since the current at any grain or grain boundary is proportional to the excess hole concentration at that point. Equating equations (4.15) and (4.8) gives the relationship between p_{1R} and p_{2L}, and substitution into equation (4.14) gives the current density at the left edge of the second grain:

$$J_{2L} = qS_{GB1}p_{2L} \qquad (4.18)$$

where

$$S_{GB1} = a_{gb} - \frac{b_{gb}^2}{a_{gb} + S_{G1}} \quad (4.19)$$

Finally, repetition of the above procedure in the second grain gives the hole current density at the right edge of the second grain:

$$J_{2R} = qS_p p_{2R} \quad (4.20)$$

where

$$S_p = a_g - \frac{b_g^2}{a_g + S_{GB1}} \quad (4.21)$$

Extension to further grain boundaries and grains is straightforward, and merely involves a repeated application of equations (4.18)–(4.21) with the appropriate values of effective recombination velocity and excess hole concentration.

The effect of the polysilicon layer on the effective recombination velocity S_p is illustrated in Figure 4.8. The effective recombination velocity can either decrease or increase with an increasing number of grains, depending on whether the lower mobility of the grain boundary or recombination at the grain boundary dominates. When the former mechanism dominates, S_p decreases, since the lower grain boundary mobility decreases the gradient of the excess hole distribution in the single-crystal emitter region. This is the situation for the

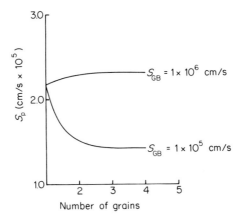

Figure 4.8. Effective recombination velocity of the polysilicon S_p as a function of the number of grains. S_p was calculated assuming the following parameter values: $L_{pp} = 0.4 \, \mu m$, $\mu_p = 103 \, cm^2/Vs$, $d_{gb} = 10 \, \text{Å}$, $\mu_{gb} = 2 \, cm^2/Vs$, $S_M = 1 \times 10^6 \, cm/s$, $d_g = 0.1 \, \mu m$ (after Yu et al. [15], copyright ©1984 IEEE)

low grain boundary recombination velocity S_{GB} of 1×10^5 cm/s in Figure 4.8. Alternatively, when the latter mechanism dominates, the additional grain boundaries give rise to extra recombination and S_p increases. This is the situation for the high grain boundary recombination velocity S_{GB} of 1×10^6 cm/s in Figure 4.8.

4.3.3 Modelling the Polysilicon/Silicon Interface

The process of depositing a polysilicon layer inevitably introduces a thin oxide-like layer at the interface between the polysilicon and single-crystal silicon. This layer is present in all types of polysilicon emitter transistor [7], although its thickness and uniformity depend critically on the type of surface treatment given prior to polysilicon deposition. For example, a wet chemical treatment such as an RCA clean gives a uniform layer approximately 14 Å thick [7], whereas an etch in hydrofluoric acid gives a discontinuous layer, varying in thickness from zero to approximately 8 Å [7]. Since silicon dioxide has a wider bandgap

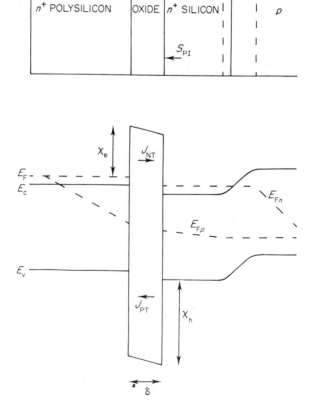

Figure 4.9. Band diagram of the polysilicon/silicon interface

than silicon, the presence of an interfacial oxide gives rise to a potential barrier to both holes and electrons.

Figure 4.9 shows a band diagram for the polysilicon/silicon interface. A rectangular potential barrier has been assumed for the interfacial oxide and the potential barriers to holes and electrons are represented by χ_h and χ_e, respectively. Carriers pass through the potential barrier mainly by tunnelling, although thermionic emission can also occur in some circumstances [16]. This latter mechanism only becomes important when the potential barrier is less than a few KT. An oxide layer would be expected to give a barrier much larger than this, and hence thermionic emission can generally be neglected.

The hole tunnelling current density J_{PT} can be expressed as [15]:

$$J_{PT} = q \left[\frac{KT}{2\pi m_h^*} \right]^{1/2} \frac{\exp - b_h}{(1 - c_h KT)} (p_{IR} - p_{IL}) \qquad (4.22)$$

where

$$b_h = \frac{4\pi\delta}{h} (2m_h^* \chi_h)^{1/2} \qquad (4.23)$$

$$c_h = \frac{2\pi\delta}{h} \left(\frac{2m_h^*}{\chi_h} \right)^{1/2} \qquad (4.24)$$

p_{IR} and p_{IL} are the hole concentrations at the right and left sides of the interfacial layer and the other symbols have their usual meaning. Equation (4.22) is the same form as equation (4.13), and can be written as:

$$J_{PT} = q T_I (p_{IR} - p_{IL}) \qquad (4.25)$$

where

$$T_I = \left[\frac{KT}{2\pi m_h^*} \right]^{1/2} \frac{\exp - b_h}{(1 - c_h KT)} \qquad (4.26)$$

From these equations it is clear that the effective recombination velocity approach used in Section 4.3.2 can be extended to include the effect of the interfacial layer.

Trapping states occur at the oxide/silicon interface in a similar way as at grain boundaries, so recombination at the interface must be included in the model. This is done in the same way as for recombination at grain boundaries by defining an interface state recombination velocity S_I:

$$S_I = \frac{N_{it}}{2} c_{pi} v_{th} \qquad (4.27)$$

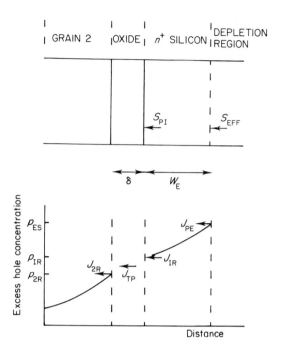

Figure 4.10. Schematic illustration of a polysilicon emitter showing the components of hole current at either side of the interfacial oxide

where N_{it} is the total density of trapping states at the interface and c_{pi} is the capture cross-section for holes at the interface. The interface traps are assumed to be equally distributed between the left and right sides of the interface.

Figure 4.10 shows a schematic illustration of the excess hole concentration in the vicinity of the interface. Following the approach of Section 4.3.2, the interface recombination velocity is defined by:

$$J_{PT} - J_{2R} = qS_I p_{2R} \qquad (4.28)$$

$$J_{IR} - J_{PT} = qS_I p_{IR} \qquad (4.29)$$

From these equations and equations (4.20) and (4.25) the current density at the right side of the interfacial layer J_{IR} can be derived in terms of the excess hole concentration at the right side of the layer p_{IR}:

$$J_{IR} = qS_{PI} p_{IR} \qquad (4.30)$$

where

$$S_{PI} = S_I + \left(\frac{1}{T_I} + \frac{1}{S_I + S_P}\right)^{-1} \qquad (4.31)$$

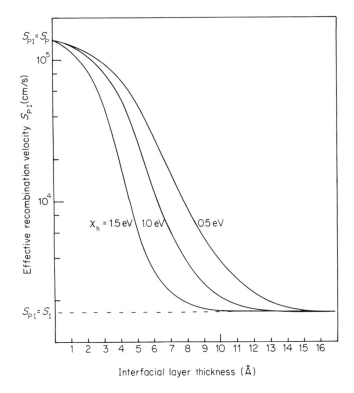

Figure 4.11. Effective recombination velocity S_{PI} of the polysilicon interface as a function of interfacial layer thickness (after Wolstenholme et al. [7] copyright American Institute of Physics 1987)

Equation (4.31), together with equations (4.21), (4.19) and (4.9), defines the effective recombination velocity at the polysilicon/silicon interface. This is represented graphically in Figure 4.11, which is a plot of S_{PI} as a function of interfacial layer thickness for three values of hole effective barrier height. A value of 1.5×10^5 cm/s has been chosen for S_P, and 1.5×10^3 for S_I [15]. Three competing mechanisms determine the value of the effective recombination velocity, namely, the transport properties of the polysilicon, tunnelling through the interfacial layer and recombination at the interface. For very thin interfacial layers, S_{PI} is approximately equal to S_P, which indicates that the transport properties of the polysilicon are dominating the transistor behaviour. Alternatively, for very thick interfacial layers, S_{PI} is approximately equal to S_I, which indicates that recombination at the interface is the dominant mechanism. In practical devices, recombination in the single-crystal emitter would also be important in this region of the characteristic, as will be described in the following section. For intermediate interface thicknesses tunnelling is the dominant mechanism, and S_{PI} varies rapidly with interfacial layer thickness.

4.3.4 Modelling the Single-crystal Emitter

In order to derive an expression for the base current of a polysilicon emitter transistor it is necessary to model recombination in the single-crystal emitter. The most effective way of achieving this is to extend the effective recombination velocity approach to the single-crystal emitter region. Unfortunately, in practical devices the emitter profile in the single-crystal emitter is non-uniform, and hence the analysis involves the integration of a function related to the doping profile. In these circumstances the use of equation (4.1) in conjunction with equation (4.31) is the simplest approach, although care must be taken for very low values of S_{PI} because equation (4.1) becomes somewhat inaccurate, as discussed in Chapter 2.

In order to arrive at a simple expression for the base current of a polysilicon emitter transistor we will assume in this section that the doping concentration in the single-crystal emitter is constant. In these circumstances extension of the effective recombination approach is straightforward and an analytical expression for the base current can be easily derived.

Following the approach of Section 4.3.2 the current density at the edge of the emitter/base depletion region can be written as:

$$J_{PE} = qS_{EFF} P_{ES} \tag{4.32}$$

where

$$S_{EFF} = a_s - \frac{b_s^2}{a_s + S_{PI}} \tag{4.33}$$

and

$$a_s = \frac{D_{pe}}{L_{pe}} \coth \frac{W_E}{L_{pe}} \tag{4.34}$$

$$b_s = \frac{D_{pe}}{L_{pe}} \operatorname{csch} \frac{W_E}{L_{pe}} \tag{4.35}$$

where W_E is the depth of the single-crystal emitter and D_{pe} and L_{pe} the hole diffusion coefficient and hole diffusion length in the single-crystal emitter. These equations are identical in form to equations (4.2)–(4.6), thereby demonstrating the ease with which the regions of a polysilicon emitter can be stacked together.

Finally, the equation for the hole current can be derived by noting that the hole concentration at the edge of the emitter/base depletion region p_{ES} is controlled by the forward bias across the emitter/base junction:

$$P_{ES} = \frac{n_i^2}{N_{DS}} \exp \frac{qV_{BE}}{KT} \tag{4.36}$$

where N_{DS} is the doping concentration in the single-crystal emitter. Substituting in equation (4.32) gives an equation for the base current of a polysilicon emitter transistor:

$$I_B = \frac{qAn_i^2}{N_{DS}} S_{EFF} \exp \frac{qV_{BE}}{KT} \quad (4.37)$$

Although this analysis has been carried out for an idealized polysilicon emitter, it is nevertheless useful for giving an insight into the operation of practical devices. It can also be used to investigate the trade-offs in polysilicon emitter design, as will be illustrated in Section 4.5.

4.4 EMITTER RESISTANCE

The band diagram in Figure 4.9 indicates that the interfacial layer gives rise to a potential barrier for both holes and electrons. We would therefore expect the electron current to be influenced in a similar way to the hole current. This is indeed the case, although the effect is only seen at high collector currents, where it shows itself as an undesirable increase in the emitter resistance of the transistor. An analytical expression for the emitter resistance can be obtained from the equation for the electron tunnelling current density in degenerately doped polysilicon [8]:

$$J_{NT} = A_e T^2 \exp - \left(\frac{E_c - E_F}{KT}\right) \frac{\exp(-b_e)}{1 - c_e KT} 2 \sinh \frac{qc_e V_o}{2} \quad (4.38)$$

where

$$b_e = \frac{4\pi\delta}{h} (2m_e^* \chi_e)^{1/2} \quad (4.39)$$

$$c_e = \frac{2\pi\delta}{h} \left(\frac{2m_e^*}{\chi_e}\right)^{1/2} \quad (4.40)$$

V_o is the voltage drop across the interfacial oxide and A_e is the modified Richardson constant, which is given by:

$$A_e = \frac{4\pi q m_e^* k^2}{h^3} \quad (4.41)$$

The emitter resistance is essentially V_o/I_{NT}, and can be calculated from equation (4.38).

In general, the emitter resistance is non-linear, as can be seen from the form of equation (4.38). This is also apparent in Figure 4.12, which shows a graph of V_o/I_C as a function of base/emitter voltage for a $6 \times 6\,\mu m^2$ polysilicon

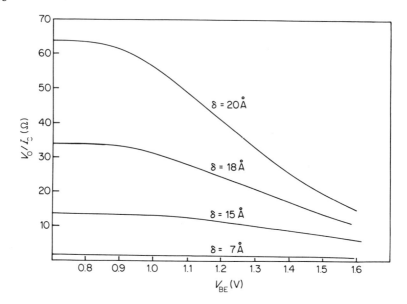

Figure 4.12. Emitter resistance of a polysilicon emitter bipolar transistor as a function of base/emitter voltage (after Ashburn *et al.* [8], copyright ©1987 IEEE)

emitter transistor. These data were computed using the BIPOLE device simulation program [20]. It is clear from this figure that the emitter resistance varies strongly with base/emitter voltage, particularly for thick interfacial oxides. However, at biases of practical interest ($V_{BE} < 0.9$ V) the emitter resistance is approximately constant, and hence can be realistically modelled by a resistor.

The value of emitter resistance at low bias can be obtained from equation (4.38) by noting that the argument of the sinh term is small with respect to unity when V_o is small. In this case, the sinh term can be approximated by its argument to give:

$$R_E = \frac{(1 - c_e KT)}{q A c_e T^2 A_e} \exp b_e \exp\left(\frac{E_c - E_F}{KT}\right) \tag{4.42}$$

The emitter resistance is therefore a function of the interfacial layer thickness δ and the electron effective barrier height χ_e through the terms b_e and c_e. It is also a function of the doping concentration in the polysilicon, through the $(E_c - E_F)$ term.

The predicted dependence of emitter resistance on interfacial oxide thickness and polysilicon doping is illustrated in Figure 4.13 for a $6 \times 6\ \mu m^2$ polysilicon emitter. It can be seen that the emitter resistance increases very rapidly at low doping concentrations and large interfacial oxide thicknesses. For this reason, interfacial oxide thicknesses greater than about 15 Å are to be avoided in practical devices.

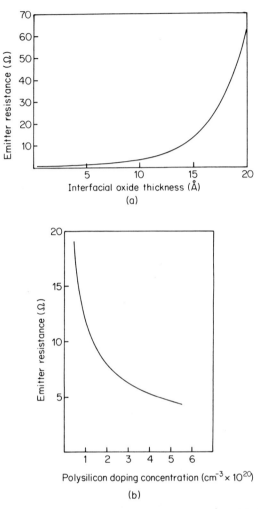

Figure 4.13. Variation of emitter resistance with (a) interfacial oxide thickness and (b) polysilicon doping concentration (after Ashburn *et al.* [8], copyright © 1987 IEEE)

4.5 DESIGN OF PRACTICAL POLYSILICON EMITTER TRANSISTORS

The most important decision that must be taken in designing a polysilicon emitter transistor is the extent that the current gain should be traded for emitter resistance. Current gains of 1000 and more can easily be realized in a device with a deliberately grown interfacial layer [7], but only at the cost of increased emitter resistance [13]. In large-geometry devices this extra emitter resistance is unlikely to cause any problems, but at sub-micron geometries it could be large enough to significantly degrade circuit performance [14]. The third important parameter of a polysilicon emitter is the emitter/base capacitance. As described earlier, the sidewall component of emitter/base capacitance can be controlled by a suitable choice of junction depth. The planar component is determined by the choice of base doping, and this will be considered in more detail in Chapters 6 and 7. These are therefore the main design criteria for a polysilicon emitter, and in this section we will consider how they can be met in practical transistors.

4.5.1 Structure of the Polysilicon/Silicon Interface

The physical structure of the polysilicon/silicon interface has been studied extensively in the literature using high-resolution transmission electron microscopy [7]. This work has shown that the surface treatment prior to polysilicon deposition, the emitter implant dose and the emitter drive-in temperature are the processing variables that have the greatest effect on the interface. Figure 4.14 illustrates the influence of the surface treatment for a device given an emitter implant of 1×10^{16} cm^{-2} and a drive-in at 900°C.

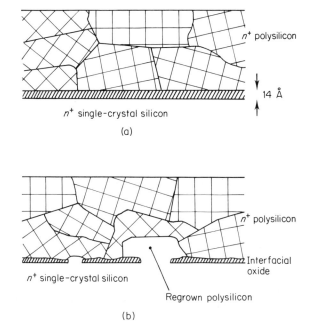

Figure 4.14. Schematic illustration of the structure of the polysilicon/silicon interface for (a) a transistor with a deliberately grown interfacial oxide (i.e. one given an RCA clean prior to polysilicon deposition) and (b) a transistor without a deliberately grown interfacial oxide (i.e. one given an HF etch prior to polysilicon deposition) (after Wolstenholme *et al.* [7], copyright American Institute of Physics 1987)

When a wet chemical surface treatment, such as an RCA clean [4], is given prior to polysilicon deposition, a continuous and uniform interfacial layer is obtained, with a thickness of typically 14 ∓ 1 Å. In contrast, when an etch in hydrofluoric acid is given, a much thinner, discontinuous interfacial layer is obtained, as shown in Figure 4.14(b). In some regions the interfacial layer is as thick as 8 Å, but in others it is thinner, and in some places there are holes. Small areas of polysilicon have regrown through the holes, extending typically 25 Å vertically into the polysilicon and also laterally over the top of the remaining interfacial oxide. More detailed experiments have shown that after polysilicon deposition the interfacial oxide is continuous, with a thickness of approximately 4 Å. The break-up of the interfacial layer therefore occurs during the emitter heat cycle at 900°C.

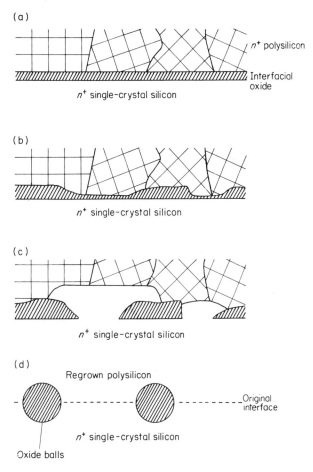

Figure 4.15. Schematic illustration of the sequence of the break-up of the interfacial oxide during a high-temperature heat-treatment (after Wolstenholme et al. [7], copyright American Institute of Physics, 1987)

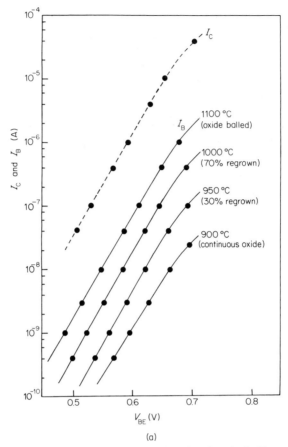

Figure 4.16. Gummel plots showing the electrical effects of the break-up of the interfacial oxide. (a) Transistors given a RCA clean prior to polysilicon deposition; (b) transistors given an HF etch prior to polysilicon deposition (after Wolstenholme et al. [7], copyright American Institute of Physics 1987)

The influence of the emitter drive-in temperature on the polysilicon/silicon interface is illustrated in Figure 4.15 for a device given an RCA clean prior to polysilicon deposition. Drive-ins at 925°C and below have no effect on the interfacial oxide, and hence a continuous, uniform layer is obtained, as shown in Figure 4.15(a). Drive-ins at temperatures of 950°C and higher lead to the break-up of the interfacial layer, with this occurring more rapidly at higher temperatures. The sequence of the break-up is illustrated in Figures 4.15(b)–(d). The first stage is a degradation of the uniformity of the interfacial layer, with some parts of the layer becoming thicker and others thinner (Figure 4.15(b)). This process continues until holes appear in the oxide, at which point the polysilicon begins to epitaxially regrow with the same orientation as the underlying silicon (Figure 4.15(c)). As the holes in the oxide become larger,

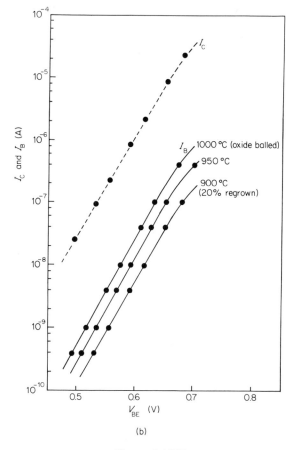

Figure 4.16(b)

the oxide thickness in the remaining areas increases (Figure 4.15(c)). Eventually the interfacial oxide breaks up completely, and the oxide forms itself into spherical balls at the interface, as illustrated in Figure 4.15(d).

The driving force for the break-up of the interfacial layer is believed to be minimization of the energy associated with the oxide layer/silicon interface [7]. In general, the energy of the interface is minimized when the surface area of the interfacial oxide is minimized. This condition is, of course, met when the oxide forms itself into spherical balls at the interface.

The rate at which the interfacial layer breaks up depends critically upon the initial thickness of the interfacial layer and also the doping concentration in the polysilicon [21]. Thinner interfacial oxides, such as those obtained using an etch in hydrofluoric acid (HF), break up more easily than thicker oxides. Therefore complete balling of an HF interfacial oxide occurs at much lower temperatures than required for an RCA interface [7]. Break-up of the interfacial oxide is also aided by the presence of dopant atoms in the polysilicon.

For example, if additional dopant is implanted into the polysilicon the interfacial oxide breaks up more rapidly during the emitter drive-in [21].

It is not surprising to discover that the break-up of the interfacial oxide has a considerable effect on the electrical characteristics of the transistors. Figure 4.16 shows Gummel plots for RCA- and HF-treated devices subjected to a short, high-temperature anneal at a temperature in the range 900–1100°C prior to the emitter implant into the polysilicon [7]. This pre-emitter anneal causes the interfacial layer to break up, with the extent of the break-up being determined by the anneal temperature. The results in Figure 4.16 show that the progressive break-up of the interfacial layer leads directly to an increase in the base current and hence a decrease in the current gain. This occurs for both RCA- and HF-treated devices, although the extent of the gain degradation is lower for the HF devices.

The above results clearly indicate that the polysilicon/silicon interface can be engineered to give different combinations of emitter resistance and current gain. For example, if high gains are of paramount importance then the process should include an RCA treatment, the polysilicon should not be too heavily doped and the emitter drive-in should be carried out at a temperature of 900°C or lower. Under these circumstances a continuous and uniform interfacial layer will result, and tunnelling of holes through the interfacial oxide will lead to suppression of the base current and hence the required high gains. Alternatively, if low emitter resistance is important the process should include an HF treatment, the polysilicon should be heavily doped and the emitter drive-in should be at a temperature of 950°C or higher. Under these circumstances the interfacial layer will be in the form of oxide balls, which will provide minimum impediment to the majority carrier flow in the emitter.

The break-up of the interfacial layer also has important consequences for device modelling and reproducibility. Clearly, a transistor with a discontinuous interfacial layer is very difficult to model in any meaningful way. In these circumstances the best approach is to define an average effective recombination velocity at the interface. The empirical approach offered by equation (4.1) is therefore likely to be more appropriate than the more rigorous theoretical approach of Section 4.3. A partially broken-up interfacial oxide is also undesirable in a production environment. Problems of batch-to-batch gain variability are likely to be encountered unless the interface treatment and the polysilicon deposition conditions are very tightly controlled. All these factors must be taken into account when designing a polysilicon emitter process.

4.5.2 Structure of the Polysilicon Layer

Transmission electron microscopy has also been used to study the structure of the polysilicon layer [7,22] under different fabrication conditions. The main process variables under the control of the device designer are polysilicon thickness, polysilicon deposition temperature, emitter implant dose, emitter dopant type and emitter drive-in conditions. All these variables have a significant effect on the structure of the polysilicon and hence to some extent on the gain of the transistor.

The major effect of the polysilicon thickness is to alter the number of grains between the interface and the emitter contact. Thin polysilicon layers, approximately 1000 Å thick, tend to have a columnar grain structure in which the grains extend completely through the layer [22], as illustrated schematically in Figure 4.17. In contrast, thicker layers generally have several grains between the interface and the emitter contact, as shown in Figure 4.14(a). For example, in a 4000 Å thick layer the grain size would be typically 1000 Å, giving, on average, four grains across the thickness of the layer. From the viewpoint of current gain, Figure 4.8 implies that a thick polysilicon layer would preferable to a thin one. However, in practice it is found that the polysilicon thickness has little effect on the current gain [11]. This is because the pseudo-grain boundary at the polysilicon/silicon interface dominates over the grain boundaries in the polysilicon. Polysilicon thickness is therefore not a critical variable in determining the gain of a polysilicon emitter transistor.

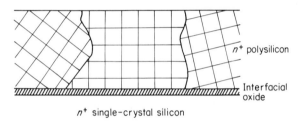

Figure 4.17. Schematic illustration of the columnar grain structure obtained for thin polysilicon layers

Polysilicon deposition temperature is important because it influences the grain structure of the polysilicon [23]. When the deposition temperature is 580°C or below, the polysilicon is deposited in the form of amorphous silicon, whereas temperatures above 580°C give polycrystalline silicon. The amorphous silicon is converted to polycrystalline form during subsequent high-temperature treatments, but the end result is larger grain sizes, lower resistances and films with a high degree of surface smoothness [23]. The lower resistance is particularly useful, since it arises from an increase in the electrically active dopant concentration in the polysilicon. This leads directly to a reduction in emitter resistance, as can be seen from Figure 4.13(b).

Dopant concentration is an important variable in determining the structure of the polysilicon. The presence of a high concentration of dopant leads to an enhancement of grain growth [24] during the emitter drive-in, the mechanism being similar to that responsible for the break-up of the interfacial oxide layer. Phosphorus is particularly effective in this role, and generally gives larger grains and lower resistances than an equivalent amount of arsenic. This leads directly to lower values of emitter resistance for phosphorus-doped polysilicon emitters [13,8] than for similar arsenic-doped devices.

4.5.3 Diffusion in Polysilicon

Arsenic diffusion in polysilicon emitters occurs in two distinct stages. The first is rapid diffusion down the grain boundaries and the second is diffusion into the bulk of the grains and, at the same time, into the single-crystal emitter. The grain boundary diffusion coefficient is several orders of magnitude higher than that in bulk silicon, and hence the arsenic penetrates to the polysilicon/silicon interface in the first few seconds of the drive-in. At this pont in time, all the arsenic is segregated at the grain boundaries [25] and the centres of the grains are undoped, as illustrated in Figure 4.18. In the remainder of the drive-in the arsenic diffuses from the grain boundaries into the bulk of the grains and through the interfacial layer into the single-crystal emitter. This process occurs at a much slower rate, and hence makes possible good control over the depth of the emitter/base junction.

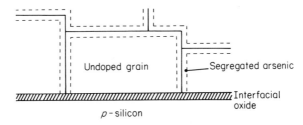

Figure 4.18. Schematic illustration of arsenic segregation in polysilicon

The emitter/base junction depth is determined by the temperature and time of the drive-in, the thickness of the interfacial layer and the arsenic concentration in the polysilicon. Several researchers have reported that the presence of an interfacial oxide retards the diffusion of arsenic into the single-crystal emitter [26,27]. Balling of the interfacial oxide and regrowth of the polysilicon would also be expected to have a dramatic effect on diffusion across the interface. Good control of the emitter/base junction depth can therefore only be achieved if the polysilicon/silicon interface is stable and under tight control. The arsenic concentration in the polysilicon also has a strong influence on the emitter/base junction depth. This comes about partly because the arsenic enhances the break-up of the interfacial oxide and partly because the grain boundaries must become saturated with arsenic before significant diffusion can occur into the single-crystal silicon. Consequently, if too little arsenic is implanted into the polysilicon the majority of it remains segregated at the grain boundaries [25] and little is left over to diffuse into the silicon.

Arsenic which is segregated at grain boundaries is generally electrically inactive, as can be seen from the doping profiles in Figure 4.19. The total arsenic concentration was measured using the SIMS technique and the electrically active arsenic concentration using Hall profiling [2]. The large peak in the total arsenic concentration at the interface is caused by segregated arsenic at the

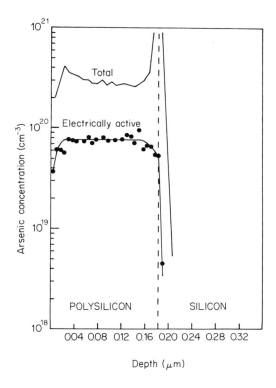

Figure 4.19. Arsenic doping profiles in a typical polysilicon emitter bipolar transistor (after Cuthbertson and Ashburn [2], copyright © 1985 IEEE)

pseudo-grain boundary between the polysilicon and silicon. This is characteristic of polysilicon emitters, and occurs for devices both with and without a deliberately grown interfacial layer [26]. The active arsenic profile is considerably different than the total arsenic profile, and indicates that only 20% of the arsenic is electrically active, the remainder being segregated at the grain boundaries. It can also be seen that the doping concentration in the polysilicon is very uniform. This simplifies the transport equations and makes possible the analysis in Section 4.3.

4.6 SIS EMITTERS

One of the reasons that polysilicon emitters are so useful in high-speed bipolar processes is that very shallow emitter/base junctions can be realized. In the limit, the junction depth of a polysilicon emitter can be reduced to zero simply by lowering the temperature of the emitter drive-in. The resulting device has no single-crystal emitter and is illustrated schematically in Figure 4.20. The electrical characteristics of this type of device bear a close resemblance to those of MIS devices [28–30], and for this reason will be referred to here as an SIS emitter.

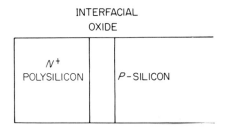

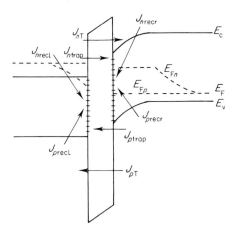

Figure 4.20. Band diagram for an SIS emitter

From the band diagram in Figure 4.20 it is clear that the physics of an SIS emitter is considerably different from that of a polysilicon emitter. In particular, a comparison with Figure 4.9 shows the presence of considerable band-bending at the surface of the *p*-type base region. Accurate modelling of this type of emitter therefore requires the inclusion of a wider range of conduction mechanisms:

(1) Direct tunnelling through the interfacial layer J_{nT} and J_{pT};
(2) Tunnelling between the valence band and surface states J_{ptrap}, and between the conduction band and surface states J_{ntrap};
(3) Shockley–Read–Hall recombination on the single-crystal side of the interfacial layer J_{precr} and J_{nrecr}.

Equations describing these mechanisms can be specified and solved numerically [31] for the base current as a function of base/emitter voltage.

Typical electrical results for practical SIS emitters [31] are shown in Figure 4.21, along with comparable results for polysilicon emitters. The major significant differences between the two types of emitter are a non-ideal base

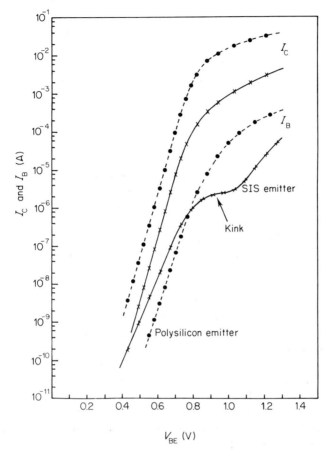

Figure 4.21. Gummel plots for an SIS emitter and a polysilicon emitter transistor (after Wolstenholme *et al.* [31])

current at low base/emitter voltages, lower values of base and collector current at high voltages and the presence of a kink in the base characteristic of the SIS device. In some cases the kink can be sufficiently pronounced that a negative resistance region is obtained [32]. These electrical characteristics of devices with SIS emitters can be understood by considering three separate ranges of forward bias [31].

At low forward biases (<0.7 V) the band-bending at the silicon surface is sufficiently large that the surface is inverted to *n*-type. In this regime the base current is dominated by recombination of holes at the silicon surface (J_{precr} in Figure 4.20). The base current is therefore sensitive to the minority carrier hole concentration at the surface and the hole recombination parameters, such as the interface state density.

As the forward bias is increased the band-bending at the silicon surface progressively reduces until the surface moves out of inversion to become *p*-type.

It is this transition of the surface from n- to p-type that defines the onset of the kink region. In this bias range ($\simeq 0.8$ V) the base current is sensitive to the minority carrier electron concentration and the electron recombination parameters, such as the interface state density.

In the kink region ($\simeq 0.9 - 1.0$ V) the gradient of the base characteristic is determined by the electron concentration at the surface. This, in turn, is controlled by two competing factors, namely, the supply of electrons from the polysilicon and the band bending at the surface. As the forward bias is increased, the supply of electrons from the polysilicon rises, which serves to enlarge the concentration of electrons at the surface. The electron tunnelling parameters, such as the effective barrier height for electrons χ_e and the interfacial layer thickness δ, are important in determining the supply of electrons from the polysilicon, and hence have an important influence on the magnitude of the kink. In opposition to this mechanism is the progressive decrease in band-bending with increasing forward bias, which causes the conduction band to move away from the electron quasi-Fermi level and thereby reduces the electron concentration. The gradient of the base characteristic can therefore either be negative or positive, depending upon which of the above mechanisms dominates. If the electron concentration at the surface is dominated by the band-bending then a negative resistance region is observed [32].

At high forward biases (> 1.1 V) the direct hole tunnelling current J_{pT}, which increases monotonically with applied bias, comes to dominate the base characteristic. The fall-off of the recombination current in the intermediate bias region (and the transition to a direct tunnelling dominated base current) is therefore the cause of the increase in the gradient of the base characteristic on the right-hand side of the kink. The base current in this high forward bias region is controlled by the hole-tunnelling parameters, such as the hole effective barrier height χ_h and the interfacial layer thickness δ.

In general, SIS emitters are inferior to polysilicon emitters because they do not provide as much collector current at a given base/emitter voltage. This can be clearly seen in the experimental characteristics of Figure 4.21. The current gain is also highly current-dependent, decreasing rapidly at low currents due to the non-ideal base characteristic. There is therefore a practical limit to the extent that the emitter/base junction depth of a polysilicon emitter transistor can be reduced, and this is determined by the onset of SIS behaviour.

References

1. J. Graul, A. Glasl and H. Murrmann, 'High-performance transistors with arsenic-implanted polysil emitters', *IEEE Jnl Solid State Circuits*, **SC11**, 491 (1976).
2. A. Cuthbertson and P. Ashburn, 'An investigation of the trade-off between enhanced gain and base doping in polysilicon emitter bipolar transistors', *IEEE Trans. Electron. Devices*, **ED32**, 2399 (1985).
3. T. H. Ning and R. D. Isaac, 'Effect of emitter contact on current gain of silicon bipolar devices', *IEEE Trans. Electron. Devices*, **ED27**, 2051 (1980).

4. H. C. De Graaff and J. G. De Groot, 'The SIS tunnel emitter: a theory for emitters with thin interfacial layers', *IEEE Trans.* **ED26**, 1771 (1979).
5. P. Ashburn and B. Soerowirdjo, 'Comparison of experimental and theoretical results on polysilicon emitter bipolar transistors', *IEEE Trans. Electron. Devices*, **ED31**, 853 (1984).
6. D. Tang and P. Solomon, 'Bipolar transistor design for optimized power-delay logic circuits', *IEEE Jnl Solid State Circuits*, **SC14**, 679 (1979).
7. G. R. Wolstenholme, N. Jorgensen, P. Ashburn and G. R. Booker, 'An investigation of the thermal stability of the interfacial oxide in polycrystalline silicon emitter bipolar transistors by comparing device results with high-resolution electron microscopy observations', *Jnl App. Phys.*, **61**, 225 (1987).
8. P. Ashburn, D. J. Roulston and C. R. Selvakumar, 'Comparison of experimental and computed results on arsenic and phosphorus doped polysilicon emitter bipolar transistors', *IEEE Trans. Electron. Devices*, **ED34**, 1346 (1987).
9. A. Neugroschel, M. Arienzo, Y. Komem and R. D. Isaac, 'Experimental study of the minority carrier transport at the polysilicon–monosilicon interface', *IEEE Trans. Electron. Devices*, **ED32**, 807 (1985).
10. G. L. Patton, J. C. Bravman and J. D. Plummer, 'Physics, technology, and modeling of polysilicon emitter contacts for VLSI bipolar transistors', *IEEE Trans. Electron. Devices*, **ED33**, 1754 (1986).
11. C. Y. Wong, C. R. M. Grovenor, P. E. Batson and D. A. Smith, 'Effect of arsenic segregation on the electrical properties of grain boundaries in polycrystalline silicon', *Jnl App. Phys.*, **57**, 438 (1985).
12. C. C. Ng and E. S. Yang, 'A thermionic diffusion model of polysilicon emitter', *IEDM Technical Digest*, 32 (1986).
13. E. F. Chor, P. Ashburn and A. Brunnschweiler, 'Emitter resistance of arsenic- and phosphorus-doped polysilicon emitter transistors', *IEEE Electron. Device Letters*, **EDL6**, 516 (1985).
14. E. F. Chor, A. Brunnschweiler and P. Ashburn, 'A propagation delay expression and its application to the optimization of polysilicon emitter ECL processes', *IEEE Jnl Solid State Circuits*, **SC23**, No. 1 (1988).
15. Z. Yu, B. Ricco and R. W. Dutton, 'A comprehensive analytical and numerical model of polysilicon emitter contacts in bipolar transistors', *IEEE Trans. Electron. Devices*, **ED31**, 773 (1984).
16. A. A. Eltoukhy and D. J. Roulston, 'The role of the interfacial layer in polysilicon emitter bipolar transistors', *IEEE Trans. Electron. Devices*, **ED29**, 1862 (1982).
17. M. M. Mandurah, K. C. Saraswat and T. I. Kamins, 'A model for conduction in polycrystalline silicon — part I: theory', *IEEE Trans. Electron. Devices*, **ED28**, 1163 (1981).
18. J. Y. W. Seto, 'The electrical properties of polycrystalline silicon films', *Jnl App. Phys.*, **46**, 5247 (1975).
19. D. M. Kim, A. N. Khondker, R. R. Shah and D. L. Crosswait, 'Conduction in polycrystalline silicon: diffusion theory and extended state mobility model', *IEEE Electron. Device Letters*, **EDL3**, 141 (1982).
20. D. J. Roulston, S. G. Chamberlain and F. Sehgal, 'Simplified computer aided analysis of double diffused transistors including two-dimensional high-level effects', *IEEE Trans. Electron. Devices*, **ED19**, 809 (1972).
21. M. C. Wilson, M. Jorgensen, G. R. Booker and P. C. Hunt, 'Electrical and microstructural investigation into the effect of arsenic emitter concentration on the enhanced gain polysilicon emitter bipolar transistor', *Proc. 5th Symposium on VLSI Technology* (1985), p. 46.
22. M. C. Wilson, P. Ashburn, B. Soerowirdjo, G. R. Booker and P. Ward, 'TEM and RBS studies of the regrowth of arsenic implanted polysilicon due to an oxidation drive-in', *Journal de Physique*, **42**, Suppl. 10, C1-253 (1982).

23. F. S. Becker, H. Oppolzer, I. Weitzel, H. Eichermuller and H. Schaber, 'Low resistance polycrystalline silicon by boron or arsenic implantation and thermal crystallization of amorphously deposited films', *Jnl App. Phys.*, **56**, 1233 (1984).
24. Y. Wada and S. Mishimatsu, 'Grain growth mechanism of heavily doped phosphorus-implanted polycrystalline silicon', *Jnl Electrochem. Soc.*, **125**, 1499 (1978).
25. M. M. Mandurah, K. C. Saraswat and C. R. Helms, 'Dopant segregation in polycrystalline silicon', *Jnl App. Phys*, **51**, 5755 (1980).
26. P. Ashburn and B. Soerowirdjo, 'Arsenic profiles in bipolar transistors with polysilicon emitters', *Solid State Electronics*, **24**, 475 (1981).
27. K. L. McLaughlin, M. A. Taylor and G. Sweeney, 'Effect of surface treatment on dopant diffusion in polycrystalline silicon capped shallow junction bipolar transistors', *App. Phys. Lett.*, **47**, 992 (1985).
28. M. A. Green, F. D. King and J. Shewchun 'Minority carrier MIS tunnel diodes and their application to electron and photovoltaic energy conversion', *Solid State Electronics*, **17**, 551 (1974).
29. M. K. Moravvej-Farshi, W. L. Guo and M. A. Green, 'Improvements in current gain and breakdown voltage of silicon MIS heterojunction emitter transistors', *IEEE Electron. Device Letters*, **EDL7**, 632 (1986).
30. M. B. Rowlandson and N. G. Tarr, 'A true polysilicon emitter transistor', *IEEE Electron. Device Letters*, **EDL6** (1985).
31. G. R. Wolstenholme, D. Browne, P. Ashburn and P. Landsberg, 'Characterization and modelling of shallow polysilicon emitter bipolar transistors', *Process and Device Modelling Conference*, Edinburgh (1986).
32. Z. Yu and R. W. Dutton, 'Gummel plot nonlinearities in polysilicon emitter transistors—including negative differential resistance behaviour', *IEEE Electron, Device Letters*, **EDL6**, 507 (1985).

Chapter 5
HETEROJUNCTION EMITTERS

5.1 INTRODUCTION

A heterojunction is a junction formed between two different semiconductor materials. The potential of heterojunction devices has long been recognized, but little progress was made towards a practical technology until the early 1970s. Since that time the emergence of molecular beam epitaxy (MBE) [1] and metal-organic chemical vapour deposition (MOCVD) [2] has resulted in revolutionary developments in this field. These techniques allow sequential layers of III/V or II/VI materials to be grown without the generation of large densities of defects at the interface. The GaAlAs/GaAs system is particularly promising in this respect because the lattice mismatch between the two semiconductors can be as small as 0.1%. Similarly, InP and GaInAsP also have near-perfect lattice match. The evolution of GaAlAs/GaAs heterojunction technology has therefore introduced a powerful new technology for the fabrication of electronic and optoelectronic circuits.

The original concept of the heterojunction emitter was proposed by Shockley in 1951 [3], and in essence is the use of a wide bandgap emitter in conjunction with a narrow bandgap base. In practice, this structure can most easily be realized by depositing a layer of GaAlAs on top of GaAs [4]. Because of the preeminence of silicon in integrated circuit technology, several attempts have also been made to fabricate silicon heterojunction transistors. The most successful approaches so far have used either semi-insulating polycrystalline silicon (SIPOS) [5] or silicon carbide [6] as the wide bandgap emitter. Working transistors have been obtained from both these approaches, but further research is required before either can be considered as a viable technology.

As with polysilicon emitters, the main advantage of heterojunction emitters is a considerably improved current gain. Although current gains above about 100 are of limited practical interest, the use of a heterojunction emitter provides additional flexibility to trade off increased gain for reduced base resistance. In this context, heterojunction bipolar transistors have been fabricated with base doping concentrations as high as $4 \times 10^{19}\,\mathrm{cm}^{-3}$ [7,8]. Further advantages accrue if the doping concentration in the emitter is decreased by several orders of magnitude to a level below that of the base. This leads directly to a significant decrease in emitter/base junction capacitance, and hence to a corresponding increase in the cut-off frequency f_T of the transistor. Emitter doping concentrations as low as 8×10^{17} cm^{-3} have been used in practice, and cut-off frequencies as high as 75 GHz achieved [9].

The design procedures for heterojunction bipolar transistors are in many ways identical to those of homojunction transistors. The only major difference is that the heterojunction designer has at his disposal an extra variable, namely the bandgap of the semiconductor. This provides additional flexibility, and subtly alters some of the trade-offs involved in designing for optimum transistor performance. In this chapter we will consider the major differences between heterojunction and homojunction transistors. Particular attention will be paid to ways in which the bandgap can be engineered to give optimum transistor performance.

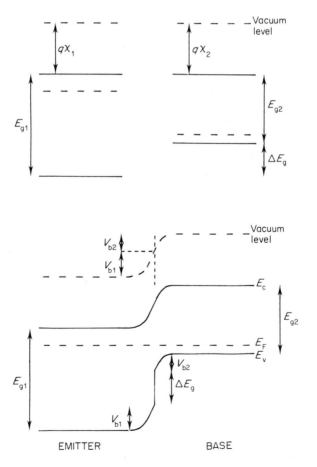

Figure 5.1. Band diagram of an idealized heterojunction emitter

5.2 THEORY OF HETEROJUNCTION EMITTERS

The basic theory of heterojunction emitters is very simple, and in a large part has already been covered in Chapter 2. Consider first Figure 5.1(a), which shows band diagrams for two semiconductors with different bandgaps E_{g1} and E_{g2}. The electron affinity, which is the energy required to remove an electron from the bottom of the conduction band to the vacuum level, is assumed in this idealized case to be the same in both semiconductors (i.e. $\chi_1 = \chi_2$). When the two semiconductors are brought together to form an abrupt heterojunction emitter, the bandgap difference ΔE_g appears in the valence band edge, as illustrated in Figure 5.1(b). In constructing a band diagram of this type it is necessary to ensure that the Fermi level is the same on both sides of the interface (in thermal equilibrium), and that the vacuum level is everywhere continuous and parallel to the band edges. The potential barrier in the valence band of a heterojunction emitter is therefore larger than for a corresponding homojunction by an amount ΔE_g, while that in the conduction band is the same. These band conditions give rise to a marked suppression of hole current but no change in the electron current. Consequently the common emitter current gain is considerably enhanced.

The improvement in current gain can be quantified by using equation (2.43) and either equation (2.44) or (2.49), depending on the value of emitter/base junction depth compared with hole diffusion length in the emitter. In most heterojunction transistors the emitter is extremely shallow, and hence equation (2.49) is the most appropriate:

$$I_C = \frac{qAD_{nb}n_{i2}^2}{W_B N_{ab}} \exp \frac{qV_{BE}}{KT} \tag{5.1}$$

$$I_B = \frac{qAD_{pe}n_{i1}^2}{W_E N_{de}} \exp \frac{qV_{BE}}{KT} \tag{5.2}$$

where n_{i1} and n_{i2} are the intrinsic carrier concentrations in the emitter and base, respectively. By analogy with the bandgap narrowing equations in Chapter 2, n_{i1} can be written as:

$$n_{i1}^2 = n_{i2}^2 \exp - \frac{\Delta E_g}{KT} \tag{5.3}$$

The common emitter current gain is therefore given by:

$$\beta = \frac{D_{nb} W_E N_{de}}{D_{pe} W_B N_{ab}} \exp \frac{\Delta E_g}{KT} \tag{5.4}$$

It can be seen that the gain of a heterojunction emitter is larger than that of a conventional emitter by the exponential factor in the above equation. In the

GaAlAs/GaAs system ΔE_g has a value of typically 0.37 eV. Insertion of this value into equation (5.4) gives a factor of approximately 10^6 improvement in gain.

The depletion layer widths and the capacitance of a heterojunction can be calculated [10] by solving Poisson's equation, subject to the boundary condition that the electric displacement is continuous at the interface: i.e.

$$\varepsilon_e E_e = \varepsilon_b E_b \tag{5.5}$$

where ε and E are the permittivity and electric field in the emitter or base. The depletion widths x_E and x_B on the emitter and base sides of the junction are given by:

$$x_E = \left[\frac{2N_{ab}\varepsilon_e\varepsilon_b(V_{bi}-V)}{qN_{de}(\varepsilon_e N_{de}+\varepsilon_b N_{ab})}\right]^{1/2} \tag{5.6}$$

$$x_B = \left[\frac{2N_{de}\varepsilon_e\varepsilon_b(V_{bi}-V)}{qN_{ab}(\varepsilon_e N_{de}+\varepsilon_b N_{ab})}\right]^{1/2} \tag{5.7}$$

where V_{bi} is the built-in voltage of the heterojunction, which is given by the sum of the partial built-in voltages V_{b1} and V_{b2}. The emitter/base junction capacitance C_{JEB} is therefore given by:

$$C_{JEB} = \left[\frac{qAN_{de}N_{ab}\varepsilon_e\varepsilon_b}{2(\varepsilon_e N_{de}+\varepsilon_b N_{ab})(V_{bi}-V)}\right]^{1/2} \tag{5.8}$$

The dielectric constant of gallium arsenide is 13.1, compared with 11.9 for silicon. The dielectric constant of GaAlAs depends upon the composition and can be calculated from:

$$\varepsilon_r = 13.1 - 3.0x \tag{5.9}$$

where x is the mole fraction of aluminium in the $Ga_{1-x}Al_xAs$. These values of dielectric constant are very similar, and indicate that for equivalent doping concentrations there is little difference between the capacitance of silicon and GaAlAs/GaAs emitters. However, as described above, considerably lower values of emitter/base capacitance can be obtained for the GaAlAs/GaAs emitter, because the doping concentration in the emitter can be reduced to a level below that of the base.

5.3 GaAlAs/GaAs HETEROJUNCTION EMITTERS

In drawing the band diagram in Figure 5.1 it was assumed that the electron affinities of the emitter and base semiconductors were the same. In general,

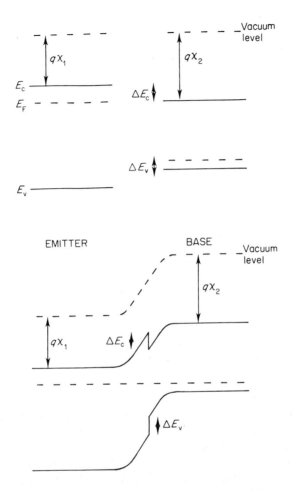

Figure 5.2. Band diagram of a GaAlAs/GaAs heterojunction emitter

however, this is not so, and in this case the idealized band diagram in Figure 5.1(b) is unrealistic. Figure 5.2 illustrates the situation for GaAlAs/GaAs heterojunctions. In this case, the difference in bandgap between the two semiconductors ΔE_g divides between the conduction and valence bands at the junction. The magnitude of the step in the conduction band ΔE_c is determined by the difference in electron affinities of the two semiconductors, and is given by:

$$\Delta E_c = q(\chi_1 - \chi_2) \tag{5.10}$$

A variety of experimental techniques have been developed [11] to measure the band discontinuities ΔE_c and ΔE_v for different heterojunctions. Initial measurements [12] for GaAlAs/GaAs heterojunctions gave $\Delta E_c/\Delta E_v = 85/15$, but more recent results [13] suggest that $\Delta E_c/\Delta E_v = 60/40$. Figure 5.3

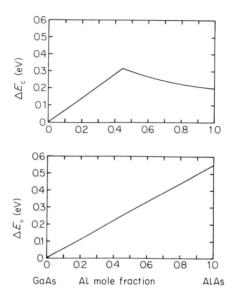

Figure 5.3. Energy band discontinuities for GaAlAs/GaAs as a function of aluminium mole fraction (after Batey and Wright [13], copyright American Institute of Physics 1986)

summarizes the most recent experimental values of ΔE_c and ΔE_v as a function of aluminium mole fraction. At the commonly used aluminium fraction of 30% Figure 5.3 gives valence and conduction band discontinuities of approximately 0.16 and 0.21 eV, respectively.

From the point of view of current gain, the band diagram of Figure 5.2 is less favourable than that of Figure 5.1. This is because the presence of the discontinuity in the conduction band reduces the emitter efficiency of the heterojunction emitter. In these circumstances the current gain can be approximated by [13]:

$$\beta = \frac{D_{nb} W_E N_{de}}{D_{pe} W_B N_{ab}} \exp \frac{\Delta E_v}{KT} \qquad (5.11)$$

Inserting the value of 0.16 eV for ΔE_v gives a factor of approximately 300 for the improvement in gain over a homojunction emitter.

In practical GaAlAs/GaAs heterojunction transistors the precise value of ΔE_v is not crucial in determining the gain. This is because other components of base current, such as recombination in the base or emitter/base depletion region, are dominant. Recombination in the emitter/base depletion region is particularly important [14] because crystallographic defects are invariably present at the interface between the two semiconductors. These arise because of mismatch, and act as generation/recombination centres. This mechanism has been considered in detail in Chapter 2, where it was shown that the base current could be approximated by equation (2.86). This equation, together with equation (5.1), then defines the current gain of a GaAlAs/GaAs heterojunction transistor. However, it must be emphasized that equation (2.86) is only approximate, and an accurate estimate of the base current requires a numerical solution [9] of equation (2.80).

Recombination in the emitter/base depletion region is also important in determining the temperature dependence of the current gain of GaAlAs/GaAs heterojunction transistors [15]. In general, the gain tends to decrease with increasing temperature for GaAlAs/GaAs heterojunction transistors, whereas for homojunction transistors it increases with temperature. The increase of gain with temperature observed in homojunction transistors is caused primarily by the higher bandgap narrowing in the emitter than in the base due to the higher emitter doping. The gain therefore increases exponentially with temperature, as predicted by equations (2.50) and (2.55). In an ideal heterojunction emitter the temperature dependence of the gain would be determined primarily by that of the base transport factor. In practice, however, this is not the case, and other mechanisms also have to be taken into account, including recombination in the emitter/base depletion region and hole injection into the emitter [15]. The net result of these competing mechanisms is a much smaller variation of gain with temperature, which, of course, is an important advantage for heterojunction transistors.

Whereas extensive measurements of bandgap narrowing have been made in heavily doping silicon, relatively few such measurements have been made in GaAs or GaAlAs. The most comprehensive data were obtained by Casey and Stern [16,17] using optical absorption measurements, and are summarized in Figure 5.4. From these data the bandgap narrowing in both p- and n-type gallium arsenide can be described by an empirical equation of the form:

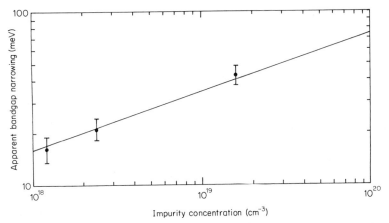

Figure 5.4. Apparent bandgap narrowing in GaAs as a function of impurity concentration (after Casey and Stern [16], copyright American Institute of Physics 1976)

$$\Delta E_g = 1.6 \times 10^{-8} (n^{1/3} + p^{1/3}) \, \text{eV} \tag{5.12}$$

In GaAlAs/GaAs heterojunction transistors the discontinuities in the band edges in Figures 5.1 and 5.2 can be eliminated by grading the composition of the boundary between the two semiconductors [18]. In this way, monotonically varying band edges can be obtained, as shown in Figure 5.5. These graded heterojunctions provide a means of eliminating the conduction band spike of Figure 5.2(b), and hence avoiding potential electrical problems [19] such as increased emitter resistance. However, the spike is not always detrimental to device performance, and in fact Kroemer [20] has postulated that it could have distinct advantages. These arise from the high energy which electrons would have to acquire in order to surmount the potential barrier represented by the spike. Electrons would therefore be injected into the base with considerable kinetic energy, and hence a high velocity [20]. If this was the case, transport of electrons across the base would be near-ballistic and hence very fast. The presence of the conduction band spike therefore offers the prospect of very short transit times and consequently very high cut-off frequencies.

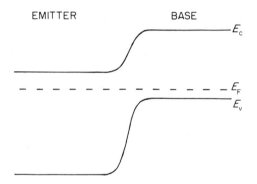

Figure 5.5. Band diagram for a heterojunction with graded composition

5.4 BANDGAP ENGINEERING

So far, we have considered the more obvious application of heterojunction technology, namely, the fabrication of heterojunction emitters for bipolar transistors. However, III/V and II/VI heterojunction technologies offer much wider possibilities than this, and in particular make it possible to engineer the bandgap to meet the required application. In this section we will therefore consider some of the alternative applications of 'bandgap engineering' as applied to bipolar devices.

As was mentioned above, the unwanted spike in the conduction band of abrupt heterojunctions can be reduced by grading the composition across the interface. Given the availability of this technology, it also becomes possible to grade the bandgap across the base [21,22], as illustrated in Figure 5.6.

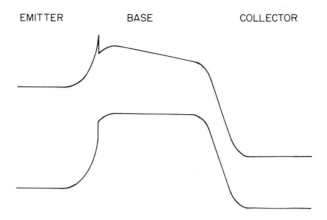

Figure 5.6. Band diagram of a heterojunction bipolar transistor with a graded base

The advantage of this structure is that it introduces a strong quasi-electric field across the base which aids the transport of minority carriers from the emitter to the collector. For a given basewidth the presence of the electric field would be expected to reduce greatly the base transit time and hence improve the cut-off frequency. Cut-off frequencies of 45 GHz have been reported [22] for GaAlAs/GaInAs/GaAs heterojunction transistors using this approach.

In homojunction transistors the forward gain is generally very much higher than the inverse one. In the majority of high-speed applications this is of no consequence, since the collector/base junction is rarely forward biased. The major exception to this rule is integrated injection logic, where the collector and emitter are essentially interchanged [23]. The very low inverse gain of the I^2L transistor is a very serious disadvantage, since it limits the fan-out obtainable from the logic gate. In heterojunction technology, however, the possibility exists of using a heterojunction collector to improve the inverse gain [24]. Such a double heterojunction transistor would have symmetrical forward and inverse gains, and hence the fan-out limitation of I^2L would be overcome. This heterojunction I^2L technology would also be considerably faster than the notoriously slow silicon version. This can be explained by the action of the heterojunction in suppressing hole injection from the base into the collector. This hole injection represents a large parasitic stored charge, and is the major reason for the relatively slow speed of silicon I^2L circuits.

A final example of the application of bandgap engineering can be found in the requirement for integrating high-speed bipolar circuits and opto-electronic devices. The double heterojunction bipolar transistor is very attractive for opto-electronic integration because it can also function as a laser. If the transistor is operated in saturation, electrons and holes can be confined in the base, where they are able to recombine to generate photons. Since the refractive index of the base is lower than that of the emitter or collector, the photons are also optically confined in the base region, with the result that laser oscillation occurs [25]. GaAlAs/GaAs [26] and InGaAsP/InP [27] heterojunction technologies have been used to produce double heterojunction structures of this type which function both as lasers and bipolar transistors.

The above examples of bandgap engineering as applied to bipolar technology serve to emphasize the very large potential of heterojunction technology. Not only does it offer a performance advantage over conventional homojunction technologies but it also makes possible new applications which cannot be met by the old technology. In this context, heterojunction technology is a promising candidate for opto-electronic integration.

References

1. A. Y. Cho and J. R. Arthur, 'Molecular beam epitaxy', *Prog. Solid State Chem.*, **10**, Pt 3, 157 (1975).
2. R. D. Dupuis, L. A. Moudy and P. D. Dapkus, 'Preparation and properties of $Ga_{1-x}Al_xAs$–GaAs heterojunctions grown by metal organic chemical vapour deposition', *Gallium Arsenide and Related Compounds 1978, Institute of Physics Conference Series*, **45**, 1 (1979).
3. W. Shockley, US patent 2,569,347 (1951).
4. W. P. Dumke, J. M. Woodall and V. L. Rideout, 'GaAs–GaAlAs heterojunction transistors for high frequency operation', *Solid State Electronics*, **15**, 1339 (1972).
5. T. Matsushita, N. Oh-uchi, H. Hahashi and H. Yamoto, 'A silicon heterojunction transistor', *App. Phys. Lett.*, **35**, 549 (1979).
6. K. Sasaki, M. M. Rahman and S. Furukawa, 'An amorphous SiC:H emitter heterojunction bipolar transistor', *IEEE Electron. Device Lett.*, **EDL6**, 311 (1985).
7. M. F. Chang, P. M. Asbeck, K. C. Wang, G. J. Sullivan and D. L. Miller, 'AlGaAs/GaAs heterojunction bipolar transistor circuits with improved high-speed performance', *Electronics Lett.*, **22**, 1173 (1986).
8. K. C. Wang, P. M. Asbeck, M. F. Chang, G. J. Sullivan and D. L. Miller, 'High-speed circuits for lightwave communication systems implemented with (AlGa)As/GaAs heterojunction bipolar transistors', *Bipolar Circuits and Technology Meeting Digest*, 142 (1987).
9. P. Ashburn, D. V. Morgan and M. J. Howes, 'A theoretical and experimental study of recombination in silicon *pn* junctions', *Solid State Electronics*, **18**, 569 (1975).
10. R. L. Anderson, 'Experiments on Ge/GaAs heterojunctions', *Solid State Electronics*, **5**, 341 (1962).
11. T. W. Hickmott, 'Electrical measurements of band discontinuities at heterostructure interfaces', in G. Bauer, F. Kuchar and H. Heinrich (eds), *Two Dimensional Systems: Physics and New Devices*, Springer-Verlag, New York (1986).
12. R. Dingle, W. Wiegmann and C. H. Henry, 'Quantum states of confined carriers in very thin $Al_xGa_{1-x}As/GaAs/Al_xGa_{1-x}As$ heterostructures', *Phys. Rev. Lett.*, **33**, 827 (1974).
13. J. Batey and S. L. Wright, 'Energy band alignment in GaAs/GaAlAs heterostructures: the dependence on alloy composition', *Jnl App. Phys.*, **59**, 200 (1986).
14. H. Ito, 'Generation–recombination current in the emitter/base junction of AlGaAs/GaAs HBTs', *Japan Jnl App. Phys.*, **25**, 1400 (1986).

15. N. Chand, R. Fischer, T. Henderson, J. Klem, W. Kopp and H. Morkoc, 'Temperature dependence of current gain in AlGaAs/GaAs heterojunction bipolar transistors', *App. Phys. Lett.*, **45**, 1086 (1984).
16. H. C. Casey and F. Stern, 'Concentration dependent absorption and spontaneous emission of heavily doped GaAs', *Jnl App. Phys.*, **47**, 631 (1976).
17. F. Stern, 'Calculated spectral dependence of gain in excited GaAs', *Jnl App. Phys.*, **47**, 5382 (1976).
18. P. M. Asbeck, D. L. Miller, R. A. Milano, J. S. Harris, G. R. Kaelin and R. Zucca, *IEDM Technical Digest*, 629 (1981).
19. A. Marty, G. Rey and J. P. Bailby, 'Electrical behaviour of an *npn* GaAlAs/GaAs heterojunction transistor', *Solid State Electronics*, **22**, 459 (1979).
20. H. Kroemer, 'Heterostructure bipolar transistors and integrated circuits', *Proc. IEEE*, **70**, 13 (1982).
21. H. Kroemer, 'Heterostructure bipolar transistors: what should we build?', *Jnl Vac. Sci. Technol.*, **B1**, 126 (1983).
22. P. M. Asbeck, M. F. Chang, K. C. Wang, G. J. Sullivan and D. L. Miller, 'GaAlAs/GaInAs/GaAs heterojunction bipolar transistor technology for sub-35ps current mode logic circuits', *Bipolar Circuits and Technology Meeting Digest*, 25 (1986).
23. K. Hart and A. Slob, 'Integrated injection logic: a new approach to LSI', *IEEE Jnl Solid State Circuits*, **SC7**, 346 (1972).
24. H. Kroemer, 'Heterostructures for everything', *Japan Jnl App. Phys.*, **20**, Suppl. 20-1, 9 (1981).
25. J. Shibata, Y. Mori, Y. Sasaki, N. Hase, H. Serizawa and T. Kajiwara, 'Fundamental characteristics of an InGaAsP/InP laser transistor', *Electronics Letters*, **21**, 98 (1985).
26. Y. Hasumi, A. Kozen, J. Temmyo and H. Asahi, 'A GaAs/AlGaAs double heterojunction device functioning as a bipolar transistor and injection laser for optoelectronic integrated circuits', *IEEE Electron. Device Lett.*, **EDL8**, 10 (1987).
27. J. Shibita, Y. Mori, Y. Sasai, N. Hase, H. Serizawa and T. Kajiwara, 'Fundamental characteristics of an InGaAsP/InP laser transistor', *Electronics Lett.*, **21**, 98 (1985).

Chapter 6
BIPOLAR INTEGRATED CIRCUIT FABRICATION

6.1 INTRODUCTION

The design of bipolar transistors is intimately interwoven with the methods used for fabrication. Any study of bipolar transistors would therefore be incomplete without consideration of the limitations imposed by the fabrication technology. Bipolar transistors are also generally incorporated into integrated circuits, and hence the transistor designer must also be aware of the restrictions imposed by the circuit configuration. In high-speed digital circuits these restrictions can be distilled into the requirement to minimize all parasitic resistances and capacitances, as will be discussed in detail in the next chapter. For analogue circuits the requirements are more subtle, and a close interaction between the circuit and process designer is needed to extract optimum performance from the product.

The design procedure at its most fundamental is the management of engineering trade-offs, and it is essential that these are dictated by the particular application. In this context the most important trade-offs are those between speed and breakdown voltage and between analogue and digital capability. High breakdown voltages imply the use of a thick, lowly doped epitaxial layer in order to give a wide collector/base depletion region. Similarly, deep junctions are implied in order to prevent premature breakdown around the periphery of the device. These requirements run counter to those for speed, where shallow junctions are needed to minimize peripheral capacitance and a thin, heavily doped epitaxial layer to suppress charge storage and base-widening effects. For analogue applications good matching between devices is of primary importance. In addition, many analogue circuits require both *npn* and *pnp* transistors to be available, as well as high-quality resistors and capacitors. Although these requirements do not run directly counter to those of a good digital process, it is often difficult to achieve the ultimate in switching speed while maintaining reproducibility of transistor and component characteristics.

From the preceding brief consideration of the trade-offs involved in the bipolar device and process design it is not surprising to discover that a wide variety of fabrication processes are currently used in industry. In this chapter we will attempt to extract the key elements of these processes and hence compile a set of bipolar process building blocks. Common yield problems associated with these building blocks will be identified and suitable techniques for troubleshooting outlined. Finally, four examples of complete processes will be given in order to illustrate how the building blocks can be combined to form different types of bipolar process. Because of the pre-eminence of silicon technology, the emphasis will be on silicon bipolar processes, although some space will also be devoted to III/V processes.

The key features of bipolar processes are illustrated in Figure 1.1, which shows a cross-section through a simple integrated circuit transistor. Emitter and base regions are clearly needed for the transistor itself as well as additional regions to make the transistor suitable for use in integrated circuits. In particular, some form of electrical isolation must be included to prevent unwanted conduction between adjacent transistors. A buried layer is desirable to reduce the collector resistance, and this necessitates the use of an epitaxial layer. Finally, a low-resistance metallization system is needed for interconnections and for making ohmic contacts to the transistor. These are the key elements of any bipolar process, and will be considered in more detail in the following sections.

6.2 BURIED LAYER AND EPITAXY

The relatively low doping concentration in the collector of a bipolar transistor (typically 1×10^{16} cm^{-3}) introduces a large collector series resistance. This can seriously degrade the electrical performance of the transistor, giving rise to a serious reduction in the current-carrying capability of the transistor and an increase in the saturation voltage. For these reasons, a buried n^+ layer is incorporated below the active device region, as shown in Figure 1.1. This provides a low-resistance path to the collector contact, thereby short-circuiting the highly resistive epitaxial collector. In some cases, an n^+ collector diffusion, connecting the collector contact and the buried layer, is also included to further reduce the collector resistance.

The buried layer is fabricated by implanting arsenic or antimony and then diffusing the relevant dopant into the substrate at a high temperature. These dopants are chosen over phosphorus because of their very low diffusion coefficient in silicon. A dry oxygen ambient is generally chosen for the drive-in in order to produce a step in the silicon surface (Figure 6.1(b)) for later alignment to the buried layer. This step comes about because the oxidation rate of heavily doped n^+ silicon is much higher than that of lightly doped silicon [1]. As a result, more silicon is consumed over the buried layer, giving rise to a small depression, as illustrated in Figure 6.1. The buried layer junction depth is determined by the requirements for a low sheet resistance (typically 20 Ω/sq or less) to minimize collector resistance, and a low surface concentration to avoid autodoping [2-4] during epitaxy. Sheet resistances lower than about 20 Ω/sq are difficult to achieve because of defect generation during drive-in and epitaxial growth [5,6].

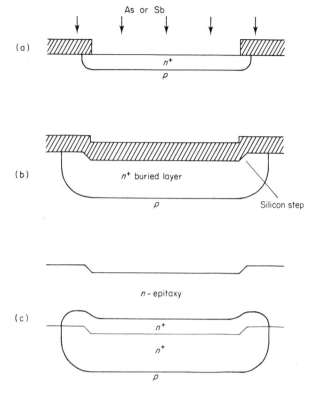

Figure 6.1. Illustration of the stages involved in the fabrication of a buried layer for a bipolar transistor

Epitaxy [1] is the term applied to the growth of a single-crystal layer of semiconductor on a single-crystal substrate. The crystalline substrate serves as a seed for the epitaxial growth, and allows the process to take place at a temperature well below the melting point of silicon. Temperatures of 950–1200°C are typically used, the lower temperatures being advantageous for thinner epitaxial layers. The epitaxial process proceeds by the reduction of a gaseous silicon compound such as silane (SiH_4), dichlorosilane (SiH_2Cl_2) or silicon tetrachloride ($SiCl_4$). Dopants can be incorporated into the growing epitaxial layer by mixing the silicon source gas with a gaseous hydride (AsH_3, PH_3 or B_2H_6). The selection of the optimum conditions for epitaxial growth is a very complex decision, based on factors such as reactor configuration, deposition temperature, growth rate, surface cleanliness, etc. The crystalline quality of epitaxial layers is of paramount importance, and is generally worse than that of the original substrate. Defects such as dislocations are able to propagate from the substrate into the growing epitaxial film, and additional defects such as epitaxial stacking faults [6] are often nucleated at impurities and damage on the substrate surface.

Epitaxial layer thickness can be controlled over a wide range of values. At the upper limit layers several hundred microns thick are routinely grown for power device applications [7], and at the lower limit layers of less than a micron are grown for high-speed digital applications [8]. In sub-micron epitaxial layers autodoping and out-diffusion of the buried layer are the main factors that constrain the extent that the thickness can be reduced. This latter mechanism is illustrated in Figure 6.1(c), where it can be seen as the diffusion of the buried layer dopant up into the epitaxial layer.

The epitaxial doping concentration can be varied from approximately 10^{20} cm^{-3} [9] down to 10^{13} cm^{-3}. These low values of doping are only achievable if the reactor is thoroughly cleaned prior to deposition. Autodoping from the rear of the wafer can be a problem, but sealing the rear surface with an oxide or nitride layer is effective in reducing this effect. In bipolar processes the epitaxial layer doping is generally determined by the requirement to suppress base-widening effects as described in Chapter 3. Epitaxial doping concentrations around 1×10^{16} cm^{-3} are typical.

Autodoping [2-4] of epitaxial layers occurs through solid state diffusion and evaporation of dopant from the substrate. The dopant is then incorporated into the growing epitaxial layer through the gas phase. This is a particular problem in bipolar processes because of the presence of the heavily doped buried layer. Its main electrical effect is an increase in the collector/base capacitance of the transistor. The epitaxial layer thickness also has to be greater than strictly necessary in order to prevent the dopant from reaching the collector junction. Autodoping can be minimized by ensuring that the buried layer surface concentration is low and by growing the epitaxial layer at a low temperature. Unwanted dopant is also incorporated into the epitaxial layer by out-diffusion from the buried layer through the interface. This is illustrated in the schematic doping profile in figure 6.2. Diffusion from the buried layer gives rise to an

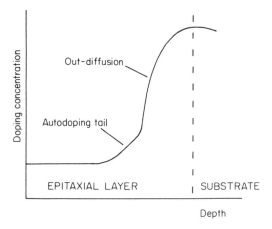

Figure 6.2. Buried layer doping profile showing the presence of out-diffusion and autodoping

approximately gaussian profile, whereas autodoping from the gas phase gives a tail at low doping concentrations.

Pattern shift and pattern distortion or washout [10–12] are also often observed in bipolar processes. Pattern shift is illustrated in Figure 6.3, where it can be seen that the depression introduced into the substrate surface during the buried layer drive-in has shifted by an amount L during epitaxy. Washout is a related but separate effect in which the shape of the pattern is distorted during epitaxy. These two phenomena make accurate alignment to the buried layer extremely difficult, particularly in high-density circuits. The magnitude of the pattern shift increases with growth rate and decreases with increasing deposition temperature. Experiments have also shown that it is significantly reduced if the epitaxy is carried out at a low pressure [13].

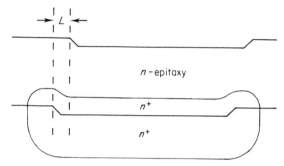

Figure 6.3. Schematic illustration of pattern-shift

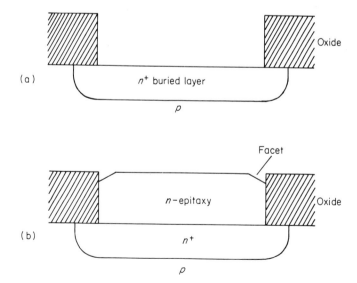

Figure 6.4. Illustration of the use of selective epitaxy to produce isolation in an integrated circuit bipolar process

Considerable advances have recently been made in techniques for selectively depositing epitaxial layers [14,15]. An example of selective epitaxy is given in Figure 6.4, where it can be seen that the silicon is deposited only in those areas where the substrate is exposed to the surface. In order to realize this structure it is necessary to suppress the deposition of silicon on the silicon dioxide regions. This is normally achieved by mixing HCl with the silicon source gas [16]. Advantages also accrue if the deposition is carried out at a reduced pressure [16]. The major problems with selective epitaxy are the formation of defects [17] and facets [18] around the periphery of the epitaxial regions. These facets are illustrated in Figure 6.4, and prevent the window from being completely filled with silicon. Experiments have shown that facet formation can be suppressed by coating the sidewall of the silicon dioxide with polysilicon [18].

6.3 ISOLATION

The simplest method of isolation for bipolar circuits is junction isolation, as illustrated in Figure 1.1. Electrical isolation between transistors or components in adjacent *n*-epitaxial islands is achieved by reverse biasing the collector/isolation *pn* junction. Since negligible current flows through a reverse-biased *pn* junction, transistors and components in adjacent wells are effectively isolated. The reverse bias is applied by connecting the substrate or isolation regions to the most negative voltage in the circuit. Although this technique is entirely effective, it suffers from the disadvantage of consuming a large amount of silicon area because of the lateral diffusion of the isolation regions. The large parasitic capacitance associated with the collector/isolation *pn* junction also makes it unsuitable for realizing high-speed circuits.

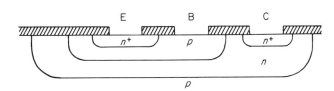

Figure 6.5. Cross-sectional view of a triple-diffused bipolar transistor

An alternative form of junction isolation is illustrated in Figure 6.5. In this case the transistor is formed using a triple diffusion into a *p*-type substrate, and isolation is provided by reverse biasing the collector/substrate *pn* junction. This approach overcomes the packing density limitation but suffers from an extremely high collector resistance. In spite of this serious drawback, this scheme is often used to provide a bipolar transistor in CMOS processes.

Oxide isolation [19] is by far the most popular method of isolation for bipolar integrated circuits. It relies on the use of a recessed silicon dioxide layer to provide isolation between adjacent islands of *n*-epi, as shown in Figure 6.6.

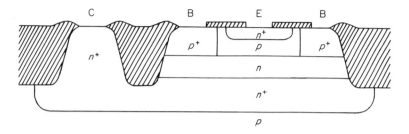

Figure 6.6. Cross-sectional view of an oxide-isolated bipolar transistor

Improved packing densities can be achieved using this approach, and the parasitic sidewall capacitance associated with junction isolation is totally eliminated. Further benefits are also obtained because the base region can be butted against the recessed oxide layer, as shown in Figure 6.6. This eliminates the sidewall component of the collector/base capacitance and also gives a further improvement in packing density. The emitter region can, in principle, also be contiguous with the recessed oxide to produce a so-called 'walled emitter' structure. This gives reduced emitter/base capacitance, but unfortunately suffers from serious yield problems due to the formation of emitter/collector pipes along the vertical oxide/silicon interface [20].

A typical fabrication sequence for oxide isolation is illustrated in Figure 6.7. The sequence starts with the growth of a thin ($\simeq 400$ Å) pad oxide and the deposition of a layer ($\simeq 1000$ Å) of silicon nitride (Figure 6.7(a)). The silicon nitride layer does not readily oxidize, and hence prevents the oxidation of the underlying silicon. The thin pad oxide relieves the stress at the nitride/silicon interface [21] and prevents the generation of dislocations during the subsequent high-temperature oxidation. Following photolithography, the epitaxial silicon is plasma etched to a depth of 55% of the required oxide thickness. This is necessary because the thickness of a silicon dioxide layer is 2.2 times that of the silicon consumed during the oxidation. The etch depth is therefore calculated to give a planar surface after the oxidation. At this stage of the process a p-type channel stop implant is often performed (Figure 6.7(b)) in order to prevent the formation of a surface inversion layer in the lightly doped substrate beneath the recessed oxide. After removal of the photoresist the wafers are oxidized to convert the exposed silicon to silicon dioxide.

The final oxide isolated structure is illustrated in Figure 6.7(c). The most notable feature of this structure is the formation of a 'bird's beak' and 'bird's head' at the periphery of the active silicon islands]22]. These are caused by lateral oxidation under the edge of the silicon nitride layer. The bird's head is undesirable because it creates an abrupt topographic feature on the surface, which can cause breaks when metallization is routed over it. The bird's beak is extremely undesirable in MOS processes because it introduces a large uncertainty in the channel width, but in bipolar processes its effect is generally

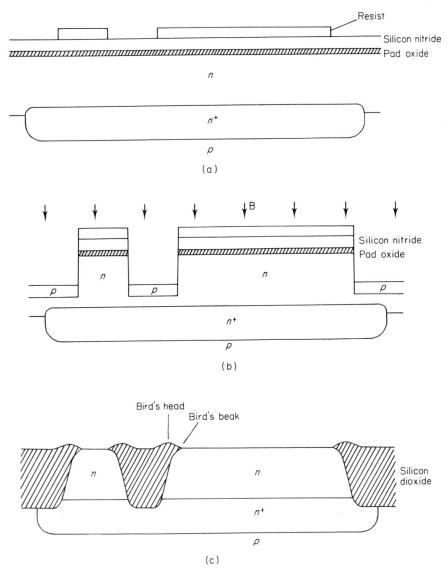

Figure 6.7. Illustration of the stages involved in the fabrication of a recessed silicon dioxide layer

not very severe, since it merely reduces the available active silicon area. Modified versions of the basic oxide isolation process have been developed which minimize bird's beak formation [23].

Further improvements in packing density can be obtained by using trench [24] or groove [25] isolation techniques, as ilustrated in Figure 6.8. Trench fabrication is a three-part process involving trench etching, refilling and planarization. The process sequence starts (Figure 6.8(a)) with the deposition

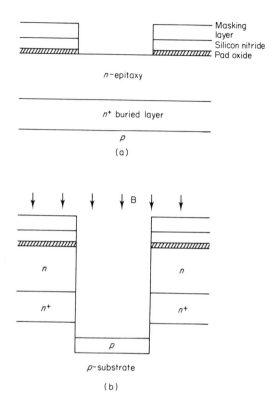

Figure 6.8. Illustration of the fabrication sequence for a trench-isolation scheme

of pad oxide, silicon nitride and a thick masking layer such as deposited silicon dioxide. The silicon nitride layer is needed as an etch stop during planarization, and the pad oxide relieves the stress at the nitride/silicon interface. Following photolithography, a deep, narrow trench is etched using reactive ion etching (Figure 6.8(b)). The main requirement for the trench is for vertical walls, a criterion that can be easily met when the trench opening is wide but which becomes more difficult to meet at sub-micron trench widths [26]. A channel-stop implant is generally introduced at this stage of the process to prevent the formation of an n-type inversion layer in the underlying lightly doped p-type silicon.

Trench refilling has been accomplished in a variety of different ways [27,28], but deposition of polysilicon is the most common. The first stage of the refill procedure is generally an oxidation to grow a thin silicon dioxide layer, as shown in Figure 6.8(c). The trench is then refilled by depositing a thick layer of undoped polysilicon. The main criterion that must be met by the refill procedure is the avoidance of defect generation during subsequent heat treatments and oxidations [29,30]. Particular problems arise if the thermal oxide around the inside of

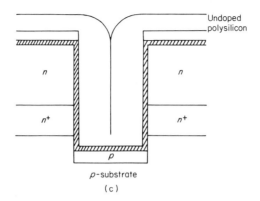

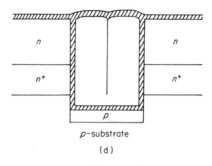

Figure 6.8

the trench is too thick, as well as at the seam in the polysilicon down the centre of the trench. When an oxidation is carried out, an oxide layer can form down the centre of the seam, forcing apart the two polysilicon layers. This generates a large amount of stress which is relieved by the formation of dislocations at the corners of the trench.

The final stage of trench isolation is planarization, which is illustrated in Figure 6.8(d). This is achieved by etching back the polysilicon to give a planar surface. The silicon nitride layer over the active transistor areas acts as an etch-stop and prevents etching of the single-crystal silicon. The trench-isolated structure is completed by carrying out an oxidation to form a capping layer over the polysilicon.

Recently, considerable interest has been shown in the use of selective epitaxy for the formation of isolation regions [31]. This technique has already been discussed in the previous section and is illustrated in Figure 6.4. It has the advantage of avoiding the formation of a bird's beak or bird's head, but problems of defect generation around the periphery of the epitaxial islands remain to be solved.

6.4 BASE

The approach used to fabricate the base of a bipolar transistor depends strongly on the required junction depth. Junctions deeper than about 0.5 μm require high-temperature diffusions to give the required penetration into the silicon, while shallow junctions can be fabricated directly using ion implantation. In this case, a low-temperature anneal is required to activate the dopant and remove the implantation damage. The temperature of the anneal needs to be high enough to completely remove the implantation damage but low enough to ensure minimal diffusion. Temperatures in the range 900–1000°C are typical. In this section we will begin by considering fabrication techniques for deep bases and then proceed to much shallower structures.

Deep bases are generally fabricated using a two-stage diffusion process [32], as illustrated in Figure 6.9. The first is a deposition stage in which the required amount of dopant is introduced into the slice. This is achieved by placing the

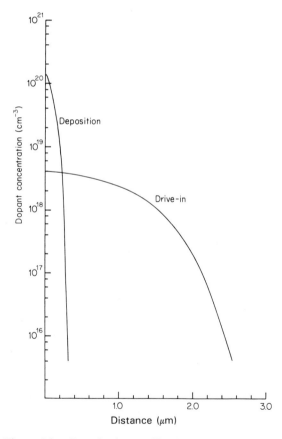

Figure 6.9. Base doping profiles for a typical analogue integrated circuit bipolar transistor after boron deposition and drive-in

water in a furnace through which an inert gas containing boron is flowing. The amount of dopant introduced depends upon the temperature, the time and the partial pressure of the boron. Deposition conditions are often arranged so that the surface concentration is at the solid solubility limit, since this gives more reproducible depositions [32]. The second stage is a high-temperature drive-in, which is necessary to lower the surface concentration and to drive the dopant deep into the silicon. This is achieved by heating the wafer in an ambient which does not contain any dopant. In practice, an oxidizing ambient is used because an oxide layer is needed over the base for the subsequent emitter fabrication.

Advantages accrue with this type of base if the chemical deposition stage is replaced by ion implantation [33]. In particular, more uniform and reproducible depositions are obtained, which in turn leads to tighter control over the base sheet resistance and the current gain. The main problem with this approach is that, unless suitable precautions are taken, dislocations are generated [33,34] when the base implantation is followed by an oxidation drive-in. The dislocations give rise to a degradation of the current gain, an increase in emitter/base and collector/base leakage currents and a decrease in the junction breakdown voltages [33]. The decrease in gain is caused by an increase in the recombination current in the emitter/base depletion region. These problems can be overcome by implanting the boron through a thin silicon dioxide layer [32] and arranging for the peak of the implant profile to occur at the oxide/silicon interface. In this situation the majority of the implantation damage occurs in the oxide layer [35], and the single-crystal silicon remains relatively undamaged.

Bases shallower than about $0.5\ \mu\mathrm{m}$ can be fabricated by implanting directly into single-crystal silicon and annealing in nitrogen at a temperature around 900°C. Problems with defect generation are again encountered if the implant is followed by an oxidation. If a silicon dioxide layer is needed for the subsequent emitter fabrication it can be produced by using low-temperature chemical vapour deposition [1] immediately after the base implant. The quality of CVD oxide is inferior to that of thermal oxide but can be significantly improved by a heat treatment at a high temperature. In this case the base anneal in nitrogen serves the dual purpose of annealing the implantation damage and densifying the silicon dioxide.

Base junctions shallower than about $0.2\ \mu\mathrm{m}$ are very difficult to realize because boron is a very light atom which penetrates significant distances into the silicon during ion implantation. It is also prone to channelling down crystal axes, which gives rise to a tail on the boron profile [35]. This can be reduced to some extent by implanting through a silicon dioxide layer [36] in order to randomize the direction of the implanted boron ions. Alternatively, the single-crystal silicon can be amorphized using a silicon implant prior to boron implantation. The problem of the large penetration depth of boron can be solved by implanting molecular boron (for example, BF_2) instead of atomic boron [37,38]. BF_2 is heavier than B^+, and hence the penetration depth is significantly reduced for a given implantation energy.

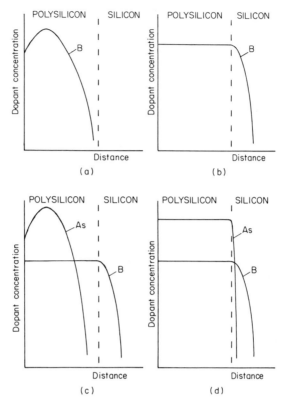

Figure 6.10. Fabrication of the base and emitter of a bipolar transistor by diffusion from polysilicon

A very simple and effective way of producing extremely shallow base junctions is to diffuse the boron from a polysilicon layer [39,40], as illustrated in Figure 6.10. Boron is implanted into a layer of undoped polysilicon, taking care to ensure that penetration into the single-crystal silicon does not occur. A short heat treatment at around 950°C is then performed to drive the boron into the silicon substrate. Junction depths of 0.15 μm and less can easily be achieved in this way. All problems of implantation induced damage are avoided, since the base is fabricated in undamaged single-crystal silicon. This approach is clearly ideal for use in conjunction with polysilicon emitters, as illustrated in Figures 6.10(c) and (d).

6.5 EMITTER

As with the base, the methods used to fabricate the emitter depend strongly on the required junction depth. Junctions of 0.5 μm and deeper are generally produced using a two-stage deposition and drive-in procedure [32]. Phosphorus is traditionally used as the emitter dopant because, unlike arsenic, it diffuses faster than boron. Control of the basewidth can therefore be achieved by adjusting the length of time of the drive-in. The deposition stage can be carried out using either a gaseous diffusion source [32] or ion implantation [41], the latter giving improved uniformity and reproducibility of the transistor characteristics. The drive-in is normally carried out in an oxidizing ambient, though better results are often obtained if an inert ambient is used [41].

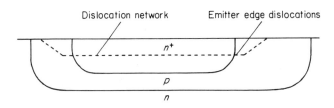

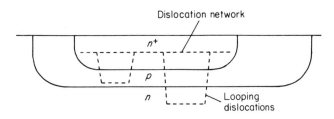

Figure 6.11. Location of dislocations in analogue bipolar integrated circuit transistors. (a) Emitter edge dislocations; (b) looping dislocations from a dislocation network (after Ashburn *et al.* [33], copyright Pergamon Press 1977)

The major problem with both ion-implanted and chemically deposited deep emitters is the generation of dislocation networks in the emitter [41-43]. This is caused by the very high phosphorus concentration [44], and is exacerbated by other factors such as implantation damage [33] and the presence of oxygen in the drive-in ambient [41]. Fortunately, the dislocation network only gives rise to a degradation of electrical characteristics when the dislocations intersect the emitter/base junction. This can occur around the periphery of the emitter [41], as shown in Figure 6.11(a), or vertically beneath the emitter [33], as illustrated in Figure 6.11(b). Dislocation networks are difficult to eliminate entirely from the emitter, but their deleterious effects can be minimized by keeping the amount of phosphorus in the emitter to a minimum [45], by eliminating implantation damage before drive-in [41] and, where possible, avoiding the use of oxygen in the drive-in ambient [41].

Emitters shallower than 0.5 μm are generally fabricated using ion implantation and an anneal at a low temperature (900–1000°C) to activate the dopant and remove the implantation damage. Arsenic is the preferred dopant because of its very low diffusion coefficient in silicon. The anneal temperature is arranged to be high enough to remove the implantation damage but not sufficiently high to give excessive diffusion. Junction depths as shallow as 0.2 μm can readily be produced in this way. As with deep phosphorus emitters, the major problem with arsenic emitters is ensuring the elimination of implantation-induced defects. This is difficult to do, and can only be achieved if the anneal is carried out in an inert ambient or an ambient containing only a small percentage of oxygen [46]. Another problem with arsenic emitters is that much of the dopant is electrically inactive [47,48]. This is largely due to precipitation of arsenic [48], and limits the maximum arsenic concentration in the emitter to around $1 \times 10^{20} \, cm^{-3}$.

The use of a polysilicon emitter overcomes all problems of implantation-induced damage [49,50]. Interactive diffusion effects, such as the emitter push effect, are also significantly reduced [51]. The fabrication sequence of polysilicon emitters is illustrated in Figures 6.10(c) and 6.10(d), where it can be seen that the emitter implant is performed into polysilicon. This has the advantage of confining all the implantation damage to the polysilicon layer. The emitter is then formed by diffusing the dopant into the undamaged, single-crystal silicon which lies below. Either inert or oxygen ambients can be used for the drive-in, and phosphorus is equally as effective as arsenic [52] as an emitter dopant. The polysilicon emitter can also be combined with a base that is diffused from polysilicon [39,40] [Figures 6.10(a) and (b)] to give a transistor that can be fabricated with extremely high yields.

6.6 YIELD PROBLEMS IN BIPOLAR PROCESSES

The major yield problem with both arsenic and phosphorus emitters is the formation of emitter/collector pipes [53]. These are localized regions of enhanced emitter diffusion which provide a resistive path between emitter and

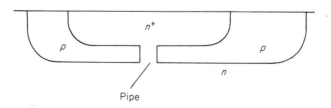

Figure 6.12. Schematic illustration of an emitter/collector pipe in a bipolar transistor

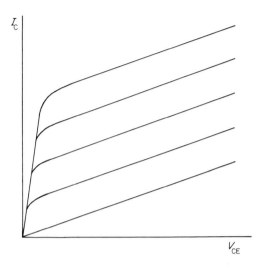

Figure 6.13. Transistor output characteristic illustrating the electrical effect of an emitter/collector pipe

collector, as illustrated schematically in Figure 6.12. Their electrical effect is shown in Figure 6.13, where it can be seen that a current flows between emitter and collector even when the base is open circuit. The resistance of the pipes can vary from several hundred ohms to a few megohms, and in severe cases, a leakage current of the order of milliamps can flow between emitter and collector. Research has shown [54,55] that pipes are caused by enhanced diffusion along defects, particularly those which intersect the emitter/base and collector/base junctions, as in Figure 6.11(b).

Other yield problems which commonly occur in bipolar processes are degradation of the current gain and increased leakage currents. The physical mechanism responsible for the increased leakage is generally the presence of generation/recombination centres in the emitter/base or collector/base depletion regions [33,41]. These recombination centres can either be caused by dislocations or other similar defects [33,41] intersecting the junctions [33,41], or alternatively by metallic contamination [32]. The most obvious symptom of recombination in the depletion region is an $\exp(qV_{BE}/mKT)$ dependence in the forward base characteristic at low currents (Figure 2.19), as discussed in Chapter 2. The reverse I/V characteristics are also often adversely effected, as illustrated in Figure 6.14. This shows typical reverse I/V characteristics for a device in which dislocations intersect the emitter/base junction and also, for comparison, a defect-free device. Of particular interest is the $I \propto V_n$ ($0.33 < n < 0.5$) dependence at low voltages observed in the device with dislocations. This type of behaviour is typical of devices dominated by generation/recombination in the depletion region.

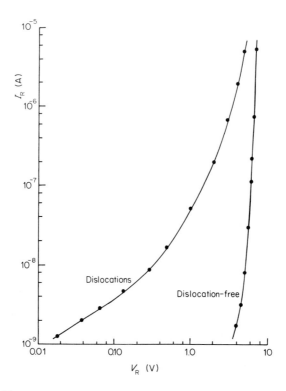

Figure 6.14. Illustration of the type of reverse current/voltage characteristics obtained when dislocations intersect a *pn* junction (after Ashburn *et al.* [33], copyright Pergamon Press 1977)

In cases where recombination in the emitter/base depletion region is suspected as being responsible for yield problems, measurements of the temperature dependence on the reverse leakage current can give additional useful information. If generation via a recombination centre is dominating the leakage, then a graph of $\ln(I_R/T^2)$ versus $1/T$ gives a straight line, the slope of which is equal to the activation energy of the dominant centre [56]. This is illustrated in Figure 6.15. In the defective device the slope is approximately equal to one half of the bandgap, indicating that the dominant centre lies close to band-centre. In contrasts, the slope obtained for the dislocation-free device is equal to the bandgap [56]. This indicates that band-to-band generation is dominating in this case.

Soft breakdown characteristics are often obtained when crystallographic defects intersect the emitter/base or collector/base depletion regions [33,6]. This can be seen in the breakdown behaviour of the defective transistor in Figure 6.14. A power law dependence of the form $I \propto V^n$ is obtained at high voltages, with *n* taking a value between 2 and 4. Metal precipitates in the depletion region can also give rise to similar characteristics [57].

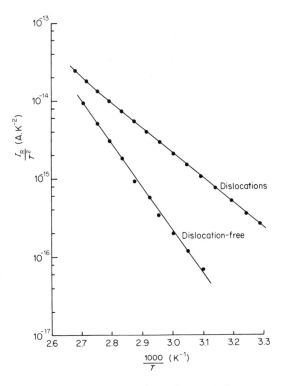

Figure 6.15. Temperature dependence of the reverse leakage current of a *pn* junction when dislocations intersect the junction (after Ashburn *et al.* [33], copyright Pergamon Press 1977)

A particular problem with phosphorus emitters is the presence of a strong interaction between the diffusion of the emitter and the base dopants [58-60]. In severe cases the base region under the emitter can be deeper than elsewhere by as much as 0.5 μm. Figure 6.16 shows a schematic illustration of this emitter push effect, and demonstrates the difficulty of controlling the basewidth under these conditions. Dissociation of phosphorus-vacancy pairs during phosphorus diffusion [61] has been identified as the mechanism that is responsible for the emitter push effect. Arsenic does not suffer from such severe diffusion

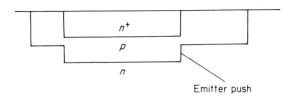

Figure 6.16. Illustration of the emitter push effect in a bipolar transistor

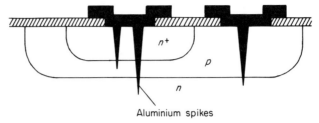

Figure 6.17. Illustration of aluminium alloy spikes in a bipolar transistor

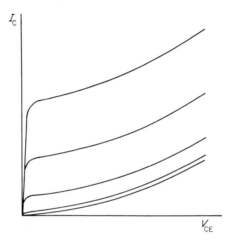

Figure 6.18. Bipolar transistor output characteristic showing the electrical effect of aluminium alloy spikes

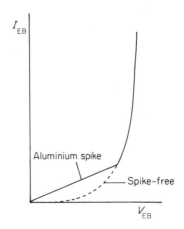

Figure 6.19. Emitter/base diode characteristic showing the electrical effect of aluminium alloy spikes

interactions as phosphorus and hence is more suitable for fabricating transistors with very narrow basewidths.

Metallization problems such as aluminium spiking [1] can cause severe yield degradation in bipolar processes, particularly where very shallow emitter/base or collector/base junctions are used. Aluminium spikes are formed by the local dissolution of silicon in the contact window during alloying. In severe cases the spikes can penetrate over 0.5 μm into the silicon, thereby shorting out the transistor junctions, as shown in Figure 6.17. The electrical effects of spikes are in many ways very similar to those of pipes, and are illustrated in Figure 6.18. When the spikes penetrate both junctions a conduction path is created between emitter and collector. However, spikes can be distinguished from pipes by noting that spikes also give rise to leaky emitter/base junctions. As a result, the gain decreases rapidly at low currents (Figure 6.18), and the emitter/base diode shows resistive behaviour at low voltages, as in Figure 6.19.

6.7 ANALOGUE BIPOLAR PROCESSES

The simplest type of bipolar process incorporates a junction isolated bipolar transistor, as illustrated in Figure 1.1. Although this process has been in production for some considerable time it is still widely used today for the design and fabrication of analogue integrated circuits. We will therefore begin our consideration of bipolar processes by studying this simple example.

The fabrication steps for the basic analogue bipolar process are summarized in Figure 6.20. Starting material of (100) or (111) orientation and approximately 5 Ωcm is typically used, and is oxidized to produce about 0.5 μm of thermal oxide. The first mask defines the buried layer, which provides a low-resistance path from the active transistor to the collector contact. Following photolithography, arsenic or antimony is ion implanted to form an n^+ region, and a high-temperature drive-in carried out to give the required sheet resistance and surface doping concentration. The resulting structure is shown in Figure 6.20(a).

Upon completion of the buried layer diffusion the silicon dioxide masking layer is removed from all parts of the wafer, thereby exposing the silicon surface. An n-type epitaxial layer, approximately 10 μm thick, is then deposited to form the collector of the transistor. During this high-temperature epitaxial stage some of the buried layer dopant diffuses up towards the surface, as illustrated in Figure 6.20(b).

Fabrication of the isolation regions begins with the growth of a new silicon dioxide layer over the entire water. The second mask is then used to define the isolation regions, which must completely surround the buried layer to form islands of n-type silicon, as shown in Figure 6.20(c). The isolation regions are formed by means of an ion implant or chemical deposition, followed by a high-temperature drive-in to push the boron completely through the epitaxial layer.

The third mask defines the base regions of the *npn* transistors and the resistors. A boron implant followed by a high-temperature drive-in is used to produce a base that is typically 3 μm deep, with a sheet resistance of 200–300 Ω/sq.

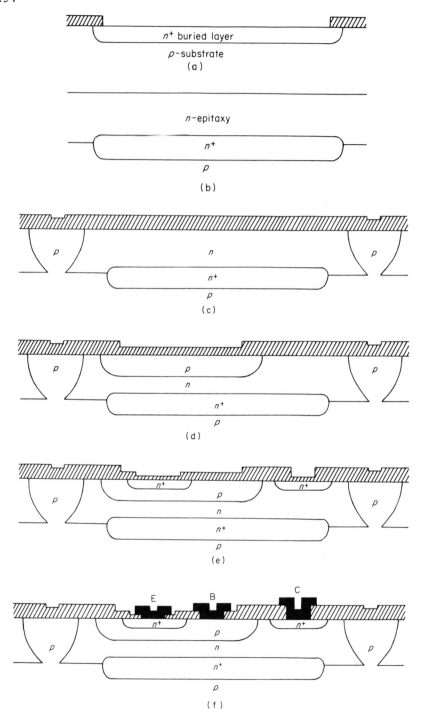

Figure 6.20. Fabrication steps for an analogue bipolar integrated circuit transistor

The drive-in is carried out in an oxidizing ambient in order to grow a thermal oxide over the base. This is needed to define an emitter window in the subsequent photolithography stage.

The fourth mask defines the emitter and collector regions, as shown in Figure 6.20(e). The n^+ diffusion underneath the collector is needed to produce a low-resistance ohmic contact to the collector. A Schottky diode would be formed at the collector if aluminium was evaporated directly onto low-doped n-type silicon. The emitter is fabricated using a phosphorus implant and a high-temperature drive-in to give a junction depth of about 2 μm and a sheet resistance of typically 20 Ω/sq. An oxygen ambient is generally used for the drive-in to provide an oxide layer for the subsequent definition of the contact windows. It is evident from figure 6.20(e) that the emitter diffusion overdopes the earlier base diffusion. This is known as compensation. There is a limit to the number of times that this can be done, since each diffusion must be more heavily doped than the last.

The final stages of the process are contact windows and metal definition. The fifth mask is used to open contact windows over the emitter, base and collector, and this is followed by the evaporation of a layer of aluminium over the entire wafer. The sixth mask then defines the aluminium interconnections between the components of the circuit. The final process step is a low-temperature (typically 450°C) alloy in a H_2/N_2 ambient, which is needed to give low-resistance ohmic contacts. The presence of the hydrogen is also important because it reduces the density of surface states at the oxide/silicon interface. These give rise to generation/recombination centres in the emitter/base

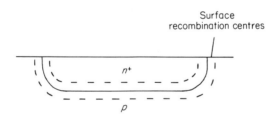

Figure 6.21. Schematic illustration showing the location of surface recombination centres in the emitter/base depletion region of a bipolar transistor

Table 6.1. Electrical parameters of a typical analogue, bipolar, integrated circuit transistor

Parameter	Typical value
β	100–150
BV_{CEO}	>15 V
BV_{CBO}	>40 V
BV_{EBO}	>6.5 V
f_T	500 MHz

depletion region at the points where it intersects the silicon surface (Figure 6.21). The alloy treatment therefore also serves the important function of improving the low current gain of the transistor. The structure of the final transistor is shown in Figure 6.20(f).

Typical electrical characteristics for the *npn* transistor are illustrated in Table 6.1. It can be seen that this type of process is suitable for circuits which operate from a 15 V supply, a requirement which is quite common in analogue applications. The cut-off frequency of 500 MHz is adequate for many applications, although more specialist bipolar processes are capable of providing much higher cut-off frequencies. In many ways the most important requirement in an analogue process is good matching between adjacent transistors and components. This is important because analogue integrated circuit design techniques often rely on the assumption that the transistors are all identical. The current mirror circuit [62] provides a simple example of this design principle.

In analogue-integrated circuit design it is important that additional components such as resistors, capacitors, diodes and *pnp* transistors are also available. These components can be fabricated without any additional processing steps using the process in Figure 6.20.

Resistors can easily be produced by using the series resistance of the various layers that comprise the bipolar transistor. Base resistors ($\simeq 200\,\Omega/\text{sq}$), emitter resistors ($\simeq 20\,\Omega/\text{sq}$) and epitaxial resistors ($\simeq 1\,\Omega/\text{sq}$) are all available. Of these, base resistors are the most useful, and are illustrated in Figure 6.22. Control of the absolute value of resistance can be obtained to about $\mp 20\%$, while matching between adjacent resistors is possible to $\mp 0.2\%$. The temperature coefficient of the base resistance is typically $\mp 15\%/°C$.

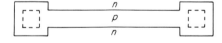

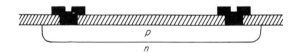

Figure 6.22. Plan and cross-sectional views of a base resistor

High-value resistors are very expensive in terms of silicon area, and hence are generally to be avoided if at all possible in integrated circuit designs. However, where these are unavoidable, pinch resistors can be used, as illustrated in Figure 6.23. In this type of resistor the base diffusion is overlaid by an emitter diffusion, thereby forcing the current to flow through the high-resistance base

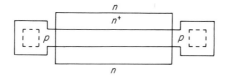

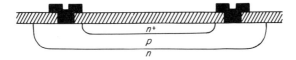

Figure 6.23. Plan and cross-sectional views of a pinch resistor

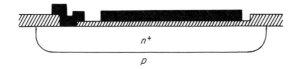

Figure 6.24. Cross-sectional view of a metal-oxide-semiconductor capacitor

region beneath the emitter. Sheet resistances of approximately 10 kΩ/sq can be achieved in this way, although matching and linearity are both significantly worse than in base resistors.

Metal oxide capacitors can easily be produced, as illustrated in Figure 6.24. This is essentially a parallel plate capacitor, the bottom plate being provided by an emitter diffusion. The capacitance per unit area is typically 50 nF/cm^2 for this type of capacitor, which demonstrates that large-value capacitors are extremely expensive in terms of silicon area.

The emitter/base and collector/base junctions of the *npn* bipolar transistor can also be used as diodes. The most common arrangement is to use the emitter/base junction as a diode and short-circuit the collector to the base. This arrangement eliminates charge storage in the collector, and hence gives very fast switching speeds. The emitter/base junction can also be used as a zener diode: sharp breakdown occurs at around 6.5 V.

PNP transistors can be produced by placing two base regions in close proximity and relying on the lateral injection of carriers to provide current gain. This arrangement is referred to as a lateral *pnp* transistor, and is illustrated in Figure 6.25. It is clear that the collector must completely surround the emitter, as shown in Figure 6.25, in order to obtain the maximum current gain. Additional improvements in gain can also be obtained if a buried layer is

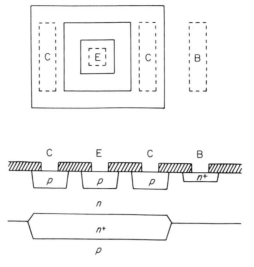

Figure 6.25. Plan and cross-sectional views of a lateral *pnp* transistor

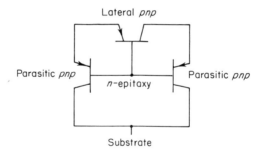

Figure 6.26. Equivalent circuit of a lateral *pnp* transistor

incorporated below the emitter of the *pnp* transistor. This has the effect of suppressing the vertical injection of holes from the emitter, thereby reducing the gain of the parasitic *pnp* transistor formed by the *p*-type emitter, the *n*-type epitaxial layer and the *p*-type substrate (Figure 6.26). Common emitter current gains of about 30 can be achieved in this way. Unfortunately, the high-frequency performance of lateral *pnp* transistors is inferior to that of vertical *npn* transistors, because of the large amount of stored charge in the epitaxial base. A cut-off frequency of around 5 MHz is typical for a lateral *pnp* transistor.

The analogue capability of the basic bipolar process in Figure 6.20 can be enhanced by including additional high-quality components in the process. This enhanced capability is, of course, obtained at the cost of an increase in process complexity and hence in cost. Matching of high-value resistors can be significantly improved if ion-implanted resistors are used (typically, 5 Ω/sq), and accurate control over the absolute value of resistance can be obtained by using laser-trimmed, metal film resistors. Additional components such as low-noise JFETs, platinum silicide Schottky diodes [63] and super β transistors [64] are also often available in specialist analogue processes.

6.8 DIGITAL BIPOLAR PROCESSES

Recent advances in silicon bipolar technology have led to a considerable improvement in digital circuit performance. Gate delays of less than 30 ps have been reported for Non-Threshold Logic (NTL) circuits, and less than 50 ps for Current Mode Logic (CML) circuits [65]. A variety of digital circuits have also been designed on high-speed bipolar processes, including frequency dividers operating at over 10 GHz, 1 K RAMs with an access time of less than 1 ns, and 6 ns 16 × 16 parallel multipliers [65]. These impressive results have been made possible by the evolution of self-aligned fabrication techniques for bipolar processes [66].

The essential features of the self-aligned bipolar process are illustrated in Figure 6.27. The fabrication sequence begins with the deposition of a layer of polysilicon over the top of the intrinsic base region of the transistor. This layer is then doped using a heavy boron implant, and a CVD silicon dioxide layer is deposited to give the structure in Figure 6.27(a). In the completed transistor the p^+ polysilicon forms the extrinsic base region of the transistor. At this point the oxide and polysilicon are patterned using reactive ion etching. It is important that this etch is highly anisotropic so that vertical walls are obtained at the edge of the window, as shown in Figure 6.27(b).

The critical stage of the self-aligned process is the formation of an oxide spacer on the sidewalls of the polysilicon. This is done by depositing a CVD oxide layer and then etching this back using reactive ion etching. An inspection of Figure 6.27(c) indicates that the deposited oxide is thicker where it covers the step in the polysilicon. The use of an anisotropic etch to remove the oxide therefore leads to the formation of a spacer on the sidewall of the polysilicon, as shown in Figure 6.27(d). The thickness of the spacer is determined by the thickness of the deposited oxide layer and by the etching characteristics of the reactive ion etcher. The transistor structure is completed by forming a polysilicon emitter and heat treating the wafers to drive the dopants from the two polysilicon layers into the single-crystal silicon (Figure 6.27(e)).

The self-aligned process gives a considerable reduction in the two most important electrical parameters of the bipolar transistor: the base resistance and the collector/base capacitance. It will be shown in Chapter 7 that these two parameters are the dominant components of the propagation delay in ECL logic circuits. The significant improvements provided by the self-aligned process therefore lead directly to a considerable improvement of circuit performance. The reduced base resistance is obtained because the p^+ extrinsic base is self-aligned to the polysilicon emitter. These two regions of the device are separated

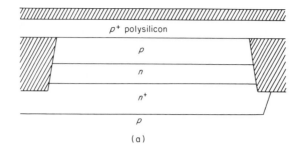

(a)

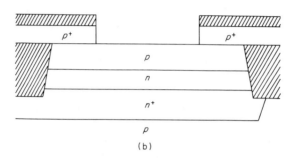

(b)

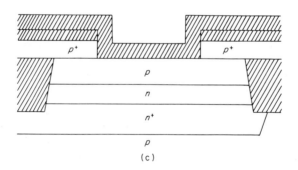

(c)

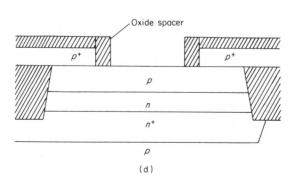

(d)

Figure 6.27

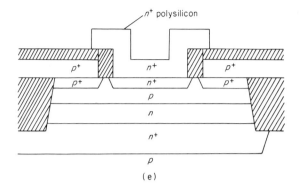

Figure 6.27. Critical fabrication steps for a self-aligned, silicon, bipolar transistor

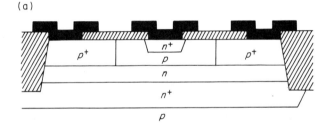

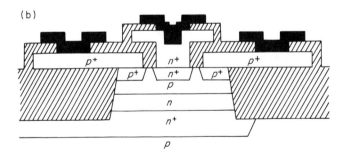

Figure 6.28. Cross-sectional views of high-speed, silicon, bipolar transistors. (a) Conventional oxide-isolated transistor; (b) self-aligned bipolar transistor

by the thickness of the oxide spacer, which is typically less than $0.4\,\mu m$ [66]. The extrinsic base region therefore extends right up to the edge of the active emitter, thereby providing a very low-resistance path to the base contact.

The reduction in collector/base capacitance can be understood by referring to Figure 6.28, which compares a conventional and a self-aligned transistor.

In the conventional transistor the size of the collector/base junction is limited by the requirement to make contact to the emitter and base. The design rules for minimum contact window size, minimum metal-to-metal separation and minimum metal overlap around the contact window therefore determine the size of the collector/base capacitance. In the self-aligned transistor contact to the base is made via the p^+ polysilicon layer, and hence there is no requirement for a contact between metal and single-crystal silicon. This means that the recessed oxide isolation regions can be brought closer together, as shown in Figure 6.28(b). The size of the extrinsic collector/base junction is then limited only by the requirement to provide an overlap between the p^+ polysilicon and the single-crystal silicon. These changes in the layout of the transistor make possible a reduction in the collector/base capacitance by as much as a factor of four.

The cross-sectional view of the self-aligned transistor in Figure 6.28(b) shows a number of additional features that are important in providing very high switching speeds. In particular, the use of a polysilicon emitter allows very shallow emitter/base junctions to be realized, thereby minimizing the peripheral component of the emitter/base capacitance. The basewidth of the transistor is arranged to be very narrow ($<0.1\,\mu$m) to minimize the base transit time. Finally, the collector doping must be high enough to suppress base-widening effects at high currents (Chapter 3). These key design features provide transistors with a cut-off frequency of over 17 GHz [67] and circuits with a gate delay of less than 30 ps [65].

In practice, the fabrication steps used to produce the oxide spacer are considerably more complicated than suggested in Figure 6.27 [68, 69]. The shape and thickness of the spacer depends critically on the conformality of step coverage during oxide deposition and on the anisotropy of the reactive ion-etching process [70]. Better results have been reported [71] if a polysilicon layer is used to form the spacer instead of a deposited oxide layer. Polysilicon gives better step coverage and can also be oxidized to produce a spacer consisting of high-quality thermal oxide. The width of the spacer is also increased, because the thickness of the final oxide layer is 2.2 times greater than that of the original polysilicon. The spacer needs to be wide enough to prevent the p^+ extrinsic base region intersecting the n^+ emitter region beneath the spacer (Figure 6.27(e)). If this occurs, an unwanted p^+n^+ junction forms around the periphery of the emitter. A large tunnelling current can flow across junctions of this type, with the result that the current gain and emitter/base breakdown voltage are significantly degraded [72].

Further enhancements to the process sequence in Figure 6.27 are necessary to provide some means of stopping the polysilicon etch at the silicon surface in Figure 6.27(b). Over-etching at this stage of the process removes dopant from the intrinsic base of the transistor and hence introduces a potential problem of batch-to-batch gain variations. Several approaches have been considered for this problem, including the use of selective etches [68] and a silicon nitride etch stop layer [69]. One simple solution is to implant the intrinsic base after

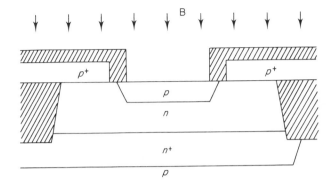

Figure 6.29. Illustration of base fabrication after oxide spacer formation

the formation of the oxide spacer, as shown in Figure 6.29. In this case, however, care must be taken to ensure that the intrinsic base joins up with the extrinsic base during the implant anneal. In practical processes a combination of the above approaches is used to produce a high-yield bipolar process with reproducible transistor characteristics.

Oxide isolation has been shown in the process in Figure 6.27, though trench isolation is also commonly used [27]. Trench isolation has the advantage of giving a reduced collector/isolation capacitance and an increased collector/isolation breakdown voltage [27]. Significantly higher packing densities can also be achieved with trench sizes of around 1 μm.

Self-aligned processing techniques have also been used to improve the performance of I^2L circuits [73,74]. Gate delays as low as 290 ps have been reported at 3 μm geometries [74]. In some processes it is possible to combine high-speed ECL circuits with high-density I^2L circuits, thereby producing a powerful technology for high-speed VLSI. Self-aligned processing techniques have also been used to improve the high-frequency performance of lateral *pnp* transistors. A cut-off frequency of 3 GHz has been reported [75] for *pnp* transistors in a process which is also capable of producing high-frequency *npn* transistors. This approach offers the prospect of high-speed digital and high-performance analogue circuits on the same chip.

6.9 GaAs/GaAlAs HETEROJUNCTION BIPOLAR PROCESSES

Although gallium arsenide fabrication techniques are very different than those of silicon, the principles of self-alignment are equally applicable to both technologies. The GaAlAs heterojunction emitter can therefore be self-aligned to the GaAs extrinsic base in a way analogous to that described above for silicon. As a result, considerable reductions of parasitic resistance and capacitance can be obtained as well as a consequent improvement in circuit performance.

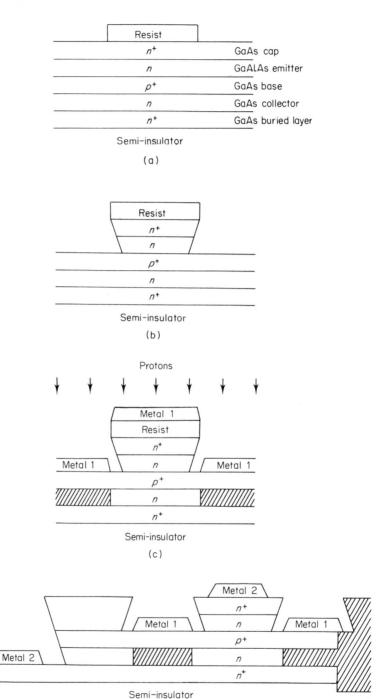

Figure 6.30. Fabrication steps for a self-aligned, AlGaAs/GaAs bipolar transistor

Table 6.2. Typical doping concentrations in a self-aligned, GaAlAs/GaAs heterojunction bipolar transistors

Layer	Thickness (μm)	Type	Doping Concentration (cm^{-3})	Al mole fraction
Cap	0.1	n^+	1×10^{19}	0
Emitter	0.2	n	8×10^{17}	0.3
Base	0.1	p^+	4×10^{19}	0
Collector	0.5	n	3×10^{16}	0
Sub-collector	0.5	n^+	5×10^{18}	0

A wide variety of self-alignment schemes have been investigated for use in GaAs/GaAlAs technology [76–80], but fortunately the simplest approach is also the most effective [78]. This is illustrated in Figure 6.30. The starting point is the deposition of the layered structure on a gallium arsenide semi-insulating substrate. Molecular beam epitaxy or metal organic chemical vapour deposition is used which gives excellent control over the layer thicknesses and doping concentrations. Typical doping concentrations for the various layers are summarized in Table 6.2. Because of the very high gains of heterojunction transistors, the base doping concentration can be extremely high, thereby avoiding the necessity for a separate extrinsic base region. Although ion implantation can be used to reduce the extrinsic base resistance still further [77], this approach is very complicated and often leads to a degradation of the yield. This occurs as a result of implantation-induced recombination centres generated during the rapid thermal annealing of the ion implant. A silicon nitride capping layer [77] must also be used during the anneal to prevent the loss of arsenic from the wafer. The n^+ layer on top of the emitter in Table 6.2 is needed to provide a low-resistance ohmic contact to the emitter.

After growth of the layered structure the emitter is defined in photoresist, as shown in Figure 6.30(a). The top layers can then be etched away to uncover the extrinsic base, as shown in Figure 6.30(b). Inward-sloping sidewalls can be produced by orienting the wafer so that the emitter stripes run parallel to the ⟨110⟩ direction [81]. If a selective etch is used [78], the etch can also be arranged to stop precisely at the emitter/base interface. With the resist still in place over the emitter, the base metal is deposited as shown in Figure 6.30(c). The separation between the base contact and the active emitter is determined by the overhang of the emitter, and is typically 0.25 μm. The base contacts are therefore self-aligned to the emitter in the same way as in silicon technology.

At this stage a reduction in the collector/base capacitance can be obtained [78] by implantation of protons [82] into the extrinsic base region, as shown in Figure 6.30(c). The lattice damage produced by the protons compensates the n-type doping, and effectively produces a recessed semi-insulating layer. Oxygen implantation [83] can also be used for this purpose. The transistor is completed by lifting off the metal from the emitter, etching via holes to the n^+ sub-collector region and depositing the emitter metal to give the structure in Figure 6.30(d).

Cut-off frequencies as high as 75 GHz have been obtained on practical, self-aligned, heterojunction transistors [84]. These transistors were produced with a base doping of 4×10^{19} cm^{-3}, a basewidth of 0.06 μm and a maximum current gain of about 40. Heterojunction technology has developed to the point that SSI circuits can also be fabricated. In particular, CML ring oscillators with gate delays of less than 30 ps have been produced, along with frequency dividers operating at over 20 GHz [84].

6.10 BICMOS PROCESSES

In recent years CMOS has emerged as the dominant technology for the design of digital VLSI circuits. Its main advantages in this application are low power consumption, large noise margin and the ease of design of CMOS circuits. However, CMOS also has a number of serious disadvantages, foremost among which is the limited ability of small-geometry MOS transistors to drive large capacitive loads. In this respect bipolar transistors are more effective than MOSFETs, because the large transconductance of the bipolar transistor gives it a greater current drive per unit silicon area. BICMOS processes allow MOS and bipolar transistors to be combined on a single chip, thereby allowing high-density MOS circuits to be combined with high current-drive bipolar circuits. Gate delays of BICMOS circuits degrade by approximately 50% in the presence of a large load capacitance, whereas the degradation for corresponding CMOS circuits is more than an order of magnitude [85]. BICMOS processes therefore offer the prospect of a considerable improvement in digital system performance [86].

BICMOS processes are also ideally suited to the implementation of analogue systems. In this context the high transconductance, high cut-off frequency and low $1/f$ noise of the bipolar transistor make it uniquely attractive for the design of many types of analogue circuit. As an example, low offset and low noise input circuits for A/D converters are much easier to design using bipolar transistors than MOSFETs. The high cut-off frequency and low circuit impedance of the bipolar transistor also brings faster settling times.

In many ways the most important advantage of BICMOS processes is that they offer a simple means of integrating a wide variety of different analogue and digital building blocks onto a single chip. This system integration approach [87] enables digital functions, such as processors and memories, to be freely integrated with analogue functions, such as A/D converters, amplifiers, filters and even transducers. In this way a powerful and universal technology is created, which makes possible for the first time the integration of all types of electronic system.

The idea of merging bipolar and MOS transistors on a single chip has been around since the 1960s [88], but little progress was made until the development of the n-well, silicon-gate CMOS process [89] in 1978. In this type of process a triple-diffused *npn* transistor can easily be fabricated, as illustrated in Figure 6.5. The collector is produced using the n-well isolation diffusion and

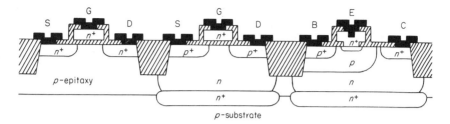

Figure 6.31. Cross-sectional view of a BICMOS bipolar transistor

the emitter using the n-channel source/drain regions of the MOS transistor [87]. An additional mask is generally needed to define the base of the bipolar transistor. Although this approach has been successfully used in a number of BICMOS processes [87,90], the resulting electrical properties of the bipolar transistor are far from optimum. The major problem is high collector resistance (typically 5 kΩ), caused by the high sheet resistance of the n-well diffusion. This limits the current-carrying capability of the bipolar transistor and precludes the use of the bipolar transistor in saturation.

Figure 6.31 shows a cross-sectional view of a fully optimized BICMOS process [85,86,91,92]. The main features of this process are the use of an n^+ buried layer to reduce the collector resistance of the bipolar transistor and a polysilicon emitter for its high gain and good yield. The requirement for an epitaxial layer in this process is, at first sight, a major disadvantage. However, many state-of-the-art CMOS processes already incorporate a low-doped epitaxial layer onto a heavily doped substrate in order to suppress latch-up [93,94]. The buried layer in Figure 6.31 therefore serves the dual function of reducing the collector resistance of the bipolar transistor and suppressing latch-up in the CMOS circuits. This latter aim is achieved by reducing the gain of the parasitic pnp transistor formed by the p-channel source/drain regions, the n-well and the p-substrate. The fully optimized BICMOS process in Figure 6.31 can be realized using three additional masking stages, one for the buried layer, one for the base and one for the polysilicon emitter.

The first stage in the fabrication of the BICMOS bipolar transistor is the implantation and diffusion of a buried layer in the usual way. This is followed by the growth of a p-type epitaxial layer approximately 5 μm thick, as shown in Figure 6.31. The n-well is fabricated using phosphorus implantation and a high-temperature drive-in to diffuse the dopant through the entire epitaxial layer. The surface concentration of the n-well must be low enough to allow the threshold voltage of the p-channel MOSFET to be controlled by a threshold implant. This means that the sheet resistance of the n-well is inevitably rather high, several kΩ/sq being typical.

Most BICMOS processes use oxide isolation, as illustrated in Figure 6.31. However, the thickness of the oxide need not be as great as in high-speed bipolar processes (Figure 6.6), since it is not necessary for it to extend through the entire thickness of the epitaxial layer. In general, therefore, the oxide isolation is optimized for the MOS transistors rather than the bipolar. The MOS transistors are very sensitive to the properties of the oxide isolation, since it defines the width of the channel. In particular, the bird's beak must be carefully controlled, since any variation in channel width leads directly to a variation in the threshold voltage.

The polysilicon emitter is fabricated in the usual way, as summarized in Figure 4.2. The complete BICMOS bipolar transistor is illustrated in Figure 6.31, where it can be seen that the n-channel source/drain implant has been used to introduce an n^+ collector region and the p-channel source/drain implant to introduce a p^+ extrinsic base. Bipolar transistors with cut-off frequencies around 4 GHz [86,91] and an excellent current-carrying capability can be produced using this approach.

For analogue applications additional components are required, such as *pnp* transistors, capacitors and resistors. If two levels of polysilicon are available in the BICMOS process, high-quality polysilicon-to-polysilicon capacitors [95] can be realized without any additional processing steps. Alternatively, MOS capacitors or metal-to-metal capacitors [96] can be used. An analogous approach can also be taken for resistor fabrication, with polysilicon resistors [97] or metal film resistors [96] being two of the most popular alternatives. Lateral *pnp* transistors can be produced using an approach analogous to that shown in Figure 6.25.

The process in Figure 6.31 is essentially a CMOS process modified to include a bipolar transistor. The alternative approach is also possible, whereby a bipolar process is modified to include MOS transistors [98,99]. This type of process is generally considerably more complicated, although it does have the advantage of being more suited to high-voltage and high-power applications [99].

References

1. S. M. Sze, *VLSI Technology*, McGraw-Hill, New York (1983).
2. G. R. Srinivasan, 'CVD epitaxial autodoping in bipolar VLSI technology', *Jnl Electrochem. Soc.*, **132**, 3005 (1985).
3. M. W. M. Graef, B. J. H. Leunissen and H. H. C. deMoor, 'Antimony, arsenic and phosphorus autodoping in silicon epitaxy', *Jnl Electrochem. Soc.*, **132**, 1942 (1985).
4. H. R. Chang, 'Autodoping in silicon epitaxy', *Jnl Electrochem. Soc.*, **132**, 219 (1985).
5. D. Robinson, G. A. Rozgonyi, T. E. Seidel and M. H. Read, 'Orientation and implantation effects on stacking faults during silicon buried layer processing', *Jnl Electrochem. Soc.*, **128**, 926 (1981).
6. K. V. Ravi, *Imperfections and Impurities in Semiconductor Silicon*, John Wiley, Chichester (1981).
7. P. D. Taylor, *Thyristor Design and Realization*, John Wiley, Chichester (1987).
8. V. J. Silvestri, G. R. Srinivasan and R. Ginsberg, 'Submicron epitaxial films', *Jnl Electrochem. Soc.*, **131**, 877 (1984).
9. J. Bloem, L. J. Giling and M. W. M. Graef, 'The incorporation of phosphorus in silicon epitaxial layer growth', *Jnl Electrochem. Soc.*, **121**, 1354 (1974).

10. S. P. Weeks, 'Pattern shift and pattern distortion during CVD epitaxy on ⟨111⟩ and ⟨100⟩ silicon', *Solid State Technol.*, **24**, 111 (1981).
11. C. M. Drum and C. A. Clark, 'Anisotropy of macrostep motion and pattern edge displacements in silicon near ⟨100⟩', *Jnl Electrochem. Soc.*, **117**, 1401 (1970).
12. P. H. Lee, M. T. Wauk, R. S. Rosler and W. C. Benzing, 'Epitaxial pattern shift comparison in vertical, horizontal and cylindrical reactor geometries', *Jnl Electrochem. Soc.*, **124**, 1824 (1977).
13. R. B. Herring, 'Advances in reduced pressure silicon epitaxy', *Solid State Technol.*, **22**, 75 (1979).
14. H. Kurten, H. J. Voss, W. Kim and W. L. Engl, 'Selective low pressure silicon epitaxy for MOS and bipolar transistor application', *IEEE Trans. Electron. Devices*, **ED30**, 151 (1983).
15. A. C. Ipri, L. Jastrzebski, J. F. Corboy and R. Metzl, 'Selective epitaxial growth for the fabrication of CMOS integrated circuits', *IEEE Trans. Electron. Devices*, **ED31**, 1741 (1984).
16. K. Tanno, N. Endo, H. Kitajima, Y. Kuroga and H. Tsuya, 'Selective silicon epitaxy using reduced pressure technique', *Japan Jnl App. Phys.*, **21**, L564 (1982).
17. H. Kitajima, A. Ishitani, N. Endo and K. Tanno, 'Crystalline defects in selectively epitaxial silicon layers', *Japan Jnl App. Phys.*, **22**, L783 (1983).
18. A. Ishitani, H. Kitajima, N. Endo and N. Kasai, 'Facet formation in selective silicon epitaxial growth', *Japan Jnl App. Phys.*, **24**, 1267 (1985).
19. J. A. Appels, E. Kooi, M. M. Paffen, J. J. H. Schlorje and W. H. C. G. Verkuylen, 'Local oxidation of silicon and its application in semiconductor technology', *Philips Res. Repts*, **25**, 118 (1970).
20. Y. Tamaki, S. Isomae, S. Mizuo and H. Higuchi, 'Evaluation of dislocation generation at silicon nitride film edges on silicon substrates by selective oxidation', *Jnl Electrochem. Soc.*, **128**, 644 (1981).
21. A. Bohg and A. K. Gaind, 'Influence of film stress and thermal oxidation on the generation of dislocations in silicon', *App. Phys. Lett.*, **33**, 895 (1978).
22. E. Bassous, H. N. Yu and V. Maniscalco, 'Topology of silicon structures with recessed silicon dioxide', *Jnl Electrochem. Soc.*, **123**, 1729 (1976).
23. K. Y. Chiu, J. L. Moll and J. Manoliu, 'A bird's beak free local oxidation technology feasible for VLSI circuits fabrication', *IEEE Trans. Electron. Devices*, **ED29**, 536 (1982).
24. S. Suyama, T. Yachi and T. Serikawa, 'A new self-aligned well isolation technique for CMOS devices', *IEEE Trans. Electron. Devices*, **ED33**, 1672 (1986).
25. A. Hayasaka, Y. Tamaki, M. Kawamura, K. Ogiue and S. Ohwaki, 'U-groove isolation technique for high speed bipolar VLSIs', *IEDM Technical Digest*, 62 (1982).
26. D. Chin, S. H. Dhong and G. J. Long, 'Structural effects on a submicron trench process', *Jnl Electrochem. Soc.*, **132**, 1705 (1985).
27. D. D. Tang, P. M. Solomon, T. H. Ning, R. D. Isaac and R. E. Burger, '1.25 micron deep groove isolated self-aligned bipolar circuits', *IEEE Jnl Solid State Circuits*, **SC17**, 925 (1982).
28. V. J. Silvestri, 'Growth kinematics of a polysilicon trench refill process', *Jnl Electrochem. Soc.*, **133**, 2374 (1986).
29. C. W. Teng, C. Slawinski and W. Hunter, 'Defect generation in trench isolation', *IEDM Technical Digest*, 586 (1984).
30. K. Sagara, Y. Tamaki and M. Kawamura, 'Evolution of dislocation generation in silicon substrates by selective oxidation of U-grooves', *Jnl Electrochem. Soc.*, **134**, 500 (1987).
31. K. A. Sabine and H. A. Kemhadjian, 'Selective epitaxy for CMOS VLSI', *IEEE Electron. Device Lett.*, **EDL6**, 43 (1985).
32. A. S. Grove, *Physics and Technology of Semiconductor Devices*, John Wiley, Chichester (1967).

33. P. Ashburn, C. Bull, K. H. Nicholas and G. R. Booker, 'Effects of dislocations in silicon transistors with implanted bases', *Solid State Electronics*, **20**, 731 (1977).
34. T. E. Seidel, R. S. Payne, R. A. Moline, W. R. Costello and J. C. C. Tsai, 'Transistors with boron bases predeposited by ion implantation and annealed in various oxygen ambients', *IEEE Trans. Electron. Devices*, **ED24**, 717 (1977).
35. J. W. Mayer, L. Eriksson and J. A. Davies, *Ion Implantation in Semiconductors*, Academic Press, New York (1970).
36. T. M. Liu and W. G. Oldham, 'Shallow boron junctions implanted in silicon through a surface oxide', *IEEE Electron. Device Lett.*, **EDL5**, 299 (1984).
37. G. Fuse, T. Hirao, K. Inoue, S. Takayanagi and Y. Yaegashi, 'Electrical properties of silicon heavily implanted with boron molecular ions', *Jnl App. Phys.*, **53**, 3650 (1982).
38. S. S. Cohen, J. F. Norton, E. F. Koch and G. J. Weisel, 'Shallow boron doped junctions in silicon', *Jnl App. Phys.*, **57**, 1200 (1985).
39. H. K. Park, K. Boyer, A. Tang, C. Clawson, S. Yu, T. Yamaguchi and J. Sachitano, 'High speed polysilicon emitter-base bipolar transistor', *IEEE Bipolar Circuits and Technology Meeting*, 39 (1986).
40. K. Kikuchi, S. Kameyama, M. Kajiyama, M. Nishio and T. Komeda, 'A high speed bipolar LSI process using self-aligned double diffusion polysilicon technology', *IEDM Technical Digest*, 420 (1986).
41. C. Bull, P. Ashburn, G. R. Booker and K. H. Nicholas, 'Effects of dislocations in silicon transistors with implanted emitters', *Solid State Electronics*, **22**, 95 (1979).
42. C. J. Bull and P. Ashburn, 'A study of diffused bipolar transistors by electron microscopy', *Solid State Electronics*, **23** 953 (1980).
43. P. Ashburn and C. J. Bull, 'Observations of dislocations and junction irregularities in bipolar transistors using the EBIC mode of the scanning electron microscope', *Solid State Electronics*, **22**, 105 (1979).
44. R. B. Fair, 'Quantified conditions for emitter misfit dislocation formation in silicon', *Jnl Electrochem. Soc.*, **125**, 923 (1978).
45. B. L. Morris and L. E. Katz, 'Reduction of excess phosphorus and elimination of defects in phosphorus emitter diffusions', *Jnl Electrochem. Soc.*, **125**, 762 (1978).
46. L. C. Parillo and B. L. Morris, 'Deleterious effects of an oxidizing drive-in ambient on implanted arsenic emitters in (111) silicon', *App. Phys. Lett.*, **35**, 345 (1979).
47. J. Murota, E. Arai, K. Kobayashi and K. Kudo, 'Relationship between total arsenic and electrically active arsenic concentrations in silicon produced by the diffusion process', *Jnl App. Phys.*, **50**, 804 (1979).
48. D. Nobili, A. Carabelas, G. Celotti and S. Solmi, 'Precipitation as the phenomenon responsible for the electrically active arsenic in silicon', *Jnl Electrochem. Soc.*, **130**, 923 (1983).
49. B. Soerowirdjo and P. Ashburn, 'Effects of surface treatments on the electrical characteristics of bipolar transistors with polysilicon emitters', *Solid State Electronics*, **26**, 495 (1983).
50. M. C. Wilson, P. Ashburn, B. Soerowirdjo, G. R. Booker and P. Ward, 'TEM and RBS studies of the regrowth of arsenic implanted polysilicon due to an oxidation drive-in', *Journal de Physique*, **43**, Suppl. 10, C1-253 (1982).
51. M. Finetti, G. Masetti, P. Negrini and S. Solmi, 'Predeposition through a polysilicon layer as a tool to reduce anomalies in phosphorus profiles and the push-out effect in npn transistors', *IEE Proc.*, Pt I, 127, 37 (1980).
52. E. F. Chor, P. Ashburn and A. Brunnschweiler, 'Emitter resistance of arsenic and phosphorus doped polysilicon emitter transistors', *IEEE Electron. Device Lett.*, **EDL6**, 516 (1985).
53. F. Barson, 'Emitter/collector shorts in bipolar devices', *IEEE Jnl Solid State Circuits*, **SC11**, 505 (1976).

54. J. P. Gowers, C. J. Bull and P. Ashburn, 'SEM and TEM observations of emitter/collector pipes in bipolar transistors', *Jnl Microscopy*, **118**, 329 (1980).
55. P. Ashburn, C. J. Bull and J. R. A. Beale, 'The use of the electron beam induced current mode of the SEM for observing emitter/collector pipes in bipolar transistors', *Jnl App. Phys.*, **50**, 3472 (1979).
56. P. Ashburn and D. V. Morgan, 'The role of radiation damage on the current-voltage characteristics of *pn* junctions', *Solid State Electronics*, **17**, 689 (1974).
57. A. Goetzberger and W. Shockley, 'Metal precipitates in silicon *pn* junctions', *Jnl App. Phys.*, **31**, 1821 (1960).
58. J. E. Lawrence, 'The cooperative diffusion effect', *Jnl App. Phys.*, **37**, 4106 (1966).
59. C. L. Jones and A. F. W. Willoughby, 'Studies of the push-out effect in silicon: I comparison of sequential boron–phosphorus and gallium–phosphorus diffusions', *Jnl Electrochem. Soc.*, **122**, 1531 (1975).
60. C. L. Jones and A. F. W. Willoughby, 'Studies of the push-out effect in silicon: II the effect of phosphorus emitter diffusion on gallium base profiles determined by radiotracer techniques', *Jnl Electrochem. Soc.*, **123**, 1531 (1976).
61. R. B. Fair, 'Effect of strain-induced bandgap narrowing on high concentration phosphorus diffusion in silicon', *Jnl App. Phys.*, **50**, 860 (1979).
62. J. Millman, *Microelectronics: Digital and Analog Circuits and Systems*, McGraw-Hill, New York (1979), p. 537.
63. S. P. Murarka, *Silicides for VLSI Applications*, Academic Press, New York (1983).
64. W. M. Gegg, J. L. Saltich, R. M. Roop and W. L. George, 'Ion implanted super-gain transistors', *IEEE Jnl Solid State Circuits*, **SC11**, 485 (1976).
65. T. Sakai, S. Konaka, Y. Yamamoto and M. Suzuki, 'Prospects of SST technology for high speed LSI', *IEDM Technical Digest*, 18 (1985).
66. T. H. Ning, R. D. Isaac, P. M. Solomon, D. D. Tang, H. N. Yu, G. C. Feth and S. K. Wiedmann, 'Self-aligned bipolar transistors for high performance and low power delay VLSI', *IEEE Trans. Electron. Devices*, **ED28**, 1010 (1981).
67. M. Suzuki, K. Hagimoto, H. Ichino and S. Konaka, 'A 9GHz frequency divider using silicon bipolar super self-aligned process technology', *IEEE Electron. Device Lett.*, **EDL6**, 181 (1985).
68. T. Sakai, Y. Yamamoto, Y. Kobayashi, K. Kawarada, Y. Inabe, T. Hayashi and H. Miyanaga, 'A 3ns 1Kbit RAM using super self-aligned process technology', *IEEE Jnl Solid State Circuits*, **SC16**, 424 (1981).
69. T. Sakai, S. Konaka, Y. Kobayashi, M. Suzuki and Y. Kawai, 'Gigabit logic bipolar technology: advanced super self-aligned process technology', *Electronics Lett.*, **19**, 283 (1983).
70. S. H. Dhong and E. J. Petrillo, 'Sidewall spacer technology for MOS and bipolar devices', *Jnl Electrochem. Soc.*, **133**, 389 (1986).
71. P. G. Kenny and P. C. Hunt, 'Composite oxide-nitride expanding sidewalls: a novel sidewall isolation technique', *Proc. ESSDERC Conf.* (1986), p. 138.
72. A. Cuthbertson and P. Ashburn, 'Self-aligned transistors with polysilicon emitters for bipolar VLSI', *IEEE Jnl Solid State Circuits*, **SC20**, 162 (1985).
73. D. D. Tang, T. H. Ning, R. D. Isaac, G. C. Feth, S. K. Wiedmann and H. N. Yu, 'Subnanosecond self-aligned I^2L/MTL circuits', *IEEE Jnl Solid State Circuits*, **SC15**, 444 (1980).
74. T. Nakamura, K. Nakazato, T. Miyazaki, T. Okabe and M. Nagata, 'High speed I^2L circuits using a sidewall base contact structure', *IEEE Trans. Electron. Devices*, **ED32**, 248 (1985).
75. K. Nakazato, T. Nakamura and M. Kato, 'A 3GHz lateral *pnp* transistor', *IEDM Technical Digest*, 416 (1986).
76. W. Dumke, J. Woodall and V. Rideout, 'GaAs/GaAlAs heterojunction transistor for high frequency operation', *Solid State Electronics*, **15**, 1339 (1972).

77. M. F. Chang, P. M. Asbeck, D. L. Miller and K. C. Wang, 'GaAs/(GaAl)As heterojunction bipolar transistors using a self-aligned substitutional emitter process', *IEEE Electron. Device Lett.*, **EDL7**, 8 (1986).
78. M. F. Chang, P. M. Asbeck, K. C. Wang, G. J. Sullivan and D. L. Miller, 'AlGaAs/GaAs heterojunction bipolar transistor circuits with improved high-speed performance', *Electronics Lett.*, **22**, 1173 (1986).
79. S. Tiwari, 'GaAlAs/GaAs heterostructure bipolar transistors: experiment and theory', *IEDM Technical Digest*, 262 (1986).
80. K. Eda, M. Inade, Y. Ota, A. Nakagawa, T. Hirosc and M. Yanagihara, 'Emitter-base-collector self-aligned heterojunction bipolar transistors using wet etching process', *IEEE Electron. Device Lett.*, **EDL7**, 694 (1986).
81. R. Fischer and H. Morkoc, 'Reduction of extrinsic base resistance in GaAs/AlGaAs heterojunction bipolar transistors and correlation with high frequency performance', *IEEE Electron. Device Lett.*, **EDL7**, 359 (1986).
82. O. Nakajima, K. Nagata, Y. Yamauchi, H. Ito and T. Ishibashi 'High-speed AlGaAs/GaAs HBTs with proton implanted buried layers', *IEDM Technical Digest*, 266 (1986).
83. P. M. Asbeck, D. L. Miller, R. J. Anderson and F. H. Eisen, 'GaAs/(GaAl)As heterojunction bipolar transistors with buried oxygen-implanted isolation layers', *IEEE Electron. Device Lett.*, **EDL5**, 310 (1984).
84. K. C. Wang, P. M. Asbeck, M. F. Chang, G. J. Sullivan and D. L. Miller, 'High speed circuits for lightwave communication systems implemented with AlGaAs/GaAs heterojunction bipolar transistors', *Bipolar Circuits and Technology Meeting Digest*, 142 (1987).
85. A. R. Alvarez, P. Meller and B. Tien '2 micron merged bipolar-CMOS technology', *IEDM Technical Digest*, 761 (1984).
86. H. Higuchi, G. Kitsukawa, T. Ikeda, Y. Nishio, N. Sasaki and K. Ogiue, 'Performance and structures of scaled-down bipolar devices merged with CMOSFETS', *IEDM Technical Digest*, 694 (1984).
87. G. Zimmer, B. Hoefflinger and J. Schneider, 'A fully implanted NMOS, CMOS, bipolar technology for VLSI of analog-digital systems', *IEEE Jnl Solid State Circuits*, **SC14**, 312 (1979).
88. H. C. Lin, J. C. Ho, R. R. Iyer and K. Kwong, 'Complementary MOS-bipolar structure', *IEEE Trans. Electron. Devices*, **ED16**, 945 (1968).
89. J. Schneider, G. Zimmer and B. Hoefflinger, 'A compatible NMOS, CMOS metal gate process', *IEEE Trans. Electron. Devices*, **ED25**, 832 (1978).
90. P. M. Zeitzoff, C. N. Anagnostopoulos, K. Y. Wong and B. P. Brandt, 'An isolated vertical *npn* transistor in an *n*-well CMOS process', *IEEE Jnl Solid State Circuits*, **SC20**, 489 (1985).
91. H. Momose, H. Shibata, S. Saitoh, J. Miyamoto, K. Kanzaki and S. Hohyama, '1.0μm *n*-well CMOS/bipolar technology', *IEEE Trans. Electron. Devices*, **ED32**, 217 (1985).
92. T. Ikeda, T. Nagano, N. Momma, K. Miyata, H. Higuchi, M. Odaka and K. Ogiue, 'Advanced BICMOS technology for high-speed VLSI', *IEDM Technical Digest*, 408 (1986).
93. J. Agraz-Guerera, R. A. Ashton, W. J. Bertram, R. C. Melin, R. C. Sun and J. T. Clemens, 'Twin-tub III: a third generation CMOS technology', *IEDM Technical Digest*, 63 (1984).
94. R. J. Smith, G. Sery, J. McCollum, J. Orton, B. Mantha, J. Smudski, T. Chi, S. Smith, J. P. Dishaw and K. Kokkonen, 'A double layer metal CHMOS III technology', *IEDM Technical Digest*, 56 (1984).
95. N. E. McGruer and R. A. Oikari, 'Polysilicon capacitor failure during rapid thermal processing', *IEEE Trans. Electron. Devices*, **ED33**, 929 (1986).

96. D. Brown, S. Chu, M. Kim, B. Gorowitz, M. Milkovic, B. Nakagawa and T. Vogelsong, 'Advanced analog CMOS technology', *IEDM Technical Digest,* 260 (1985).
97. Y. Amemiya, T. Ono and K. Kato, 'Electrical trimming of heavily doped polycrystalline silicon resistors', *IEEE Trans. Electron. Devices*, **ED26**, 1738 (1980).
98. Y. Okada, K. Kaneko, S. Kudo, K. Yamazaki and T. Okabe, 'An advanced bipolar-MOS-I^2L technology with a thin epitaxial layer for analog–digital VLSI', *IEEE Trans. Electron. Devices*, **ED32**, 232 (1985).
99. A. Anderini, C. Contiero and P. Galbiati, 'A new integrated silicon gate technology combining bipolar linear, CMOS logic, and DMOS power parts', *IEEE Trans. Electron. Devices*, **ED33**, 2025 (1986).

Chapter 7

OPTIMIZATION OF HIGH-SPEED BIPOLAR PROCESSES

7.1 INTRODUCTION

Optimization of high-speed bipolar processes is an extremely complicated task involving interactions between device, process and circuit design. The first problem is identifying an electrical parameter that can serve as a figure of merit for the process performance. Ideally, this parameter should be applicable to all types of circuit, both analogue and digital, and should also be easily measured. Unfortunately, there is no single figure of merit that satisfies these requirements, and in practice a variety of parameters have been used.

The cut-off frequency (f_T) of a bipolar transistor is often quoted as a figure of merit for a bipolar process. As discussed in Chapter 3, this is related to the forward transit time, which is a measure of the fundamental switching speed of the bipolar transistor itself. Unfortunately, the cut-off frequency f_T takes no account of parasitic resistances and capacitances associated with the transistor, and hence is a poor indicator of circuit performance. In order to overcome this limitation an alternative figure of merit f_{MAX} is often quoted. This has been described in Chapter 3 (equation (3.56)), and includes the effects of base resistance and collector capacitance. This is a more realistic figure of merit for circuit performance, although it is questionable whether it is accurate for integrated circuit bipolar transistors [1].

The propagation delay of a logic gate is by far the most reliable figure of merit for digital circuit performance. Unfortunately, this cannot be measured on a single transistor but requires the fabrication of circuits. Ring oscillators are often used for this purpose, and provide a very simple means of measuring the gate delay. A ring oscillator consists of an odd number of inverters connected together in a ring. On application of power the circuit begins to oscillate at a frequency equal to twice the propagation delay. This measurement provides an accurate and reliable figure of merit for logic gate performance, but additional factors need to be taken into account in assessing digital system performance. In particular, load capacitance and fan-out lead to a degradation of the gate delay by a factor of approximately 1.5 [2]. In MOS circuits an even greater degradation is obtained by a factor of about 4.5 [2].

Having selected the propagation delay as a suitable figure of merit we next need to devise some means of relating this to the processing parameters. Ideally, we would like to have available an analytical expression for the propagation delay in terms of the processing parameters, such as base implantation dose, emitter/base junction depth, etc. Unfortunately, this goal is difficult to achieve, and in practice the best that can be done is to compute a quasi-analytical expression for the gate delay in terms of all the time constants of the circuit [3]. These time constants are expressed in terms of transistor parameters, and can be related to the processing parameters through the modelling equations in Chapter 3 or through process and device modelling programs such as SUPREM [4] and BIPOLE [5]. This approach provides an effective means of optimizing the process for speed, since the dominant components of the gate delay can be easily identified.

In this chapter we will discuss a quasi-analytical expression for the propagation delay of an ECL gate [3]. This will then be used to calculate the propagation delays of conventional and self-aligned bipolar processes. In this way it will be demonstrated how the propagation delay expression can be used for process optimization.

7.2 ECL PROPAGATION DELAY EXPRESSION

Research has shown [6,7] that the propagation delay of bipolar logic circuits can be expressed as a linear combination of the time constants of the circuit, with each time constant weighted by a factor that is dependent upon the circuit configuration. In general, the propagation delay can therefore be approximated by:

$$\tau_d = \sum_i K_i R_i C_i + \sum_j K_j \tau_j \qquad (7.1)$$

where K_i and K_j are the weighting factors for the corresponding product terms. The R_i and C_i terms include the resistances and capacitances of the particular

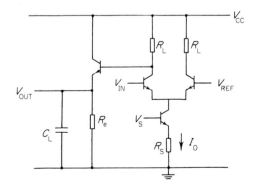

Figure 7.1. Circuit diagram of an ECL logic gate

logic circuit and the parasitic resistances and capacitances of the bipolar transistor. The τ_j terms are the forward and reverse transit times of the transistor.

For the ECL circuit in Figure 7.1 the R_i terms include the intrinsic and extrinsic base resistances R_{BI} and R_{BX}, the emitter and collector resistances R_E and R_C, the emitter follower resistance R_e and the load resistance R_L. Similarly, the C_i terms include the intrinsic and extrinsic collector/base capacitances C_{JCI} and C_{JCX}, the emitter/base capacitance C_{JEB}, the diffusion capacitance C_{DE}, the substrate capacitance C_{JCS} and the interconnection capacitance C_L. Each of the R_iC_i time constants has a weighting factor K_i associated with it, which can be computed using a sensitivity analysis [3]. The resulting expression for the propagation delay is:

$$\begin{aligned}\tau_d = &\; K_1 \tau_F \\ &+ R_{BI}(K_2 C_{JCI} + K_3 C_{JCX} + K_4 C_{JEB} + K_5 C_{DE}) \\ &+ R_{BX}(K_6 C_{JCI} + K_7 C_{JCX} + K_8 C_{JEB} + K_9 C_{DE}) \\ &+ R_L(K_{10} C_{JCI} + K_{11} C_{JCX} + K_{12} C_{JEB} + K_{13} C_{JCS} + K_{14} C_L) \\ &+ R_C(K_{15} C_{JCI} + K_{16} C_{JCX} + K_{17} C_{DE} + K_{18} C_{JCS}) \\ &+ R_E(K_{19} C_{JCI} + K_{20} C_{JCX} + K_{21} C_{JEB} + K_{22} C_{DE} + K_{23} C_{JCS} + K_{24} C_L) \end{aligned} \quad (7.2)$$

Table 7.1. Summary of the weighting factors for the ECL propagation delay expression (equation (7.2)) (after Chor et al. [3], copyright © 1988 IEEE)

i	Time constant R_iC_i	Weighting factor K_i	i	Time constant R_iC_i	Weighting factor K_i
1	τ_F	0.81 + 0.39F	13	$R_L C_{JCS}$	0.16
2	$R_{BI} C_{JCI}$	3.41	14	$R_L C_L$	0.30F
3	$R_{BI} C_{JCX}$	0.42	15	$R_C C_{JCI}$	2.65
4	$R_{BI} C_{JEB}$	0.74	16	$R_C C_{JCX}$	1.67
5	$R_{BI} C_{DE}$	0.97	17	$R_C C_{DE}$	0.32
6	$R_{BX} C_{JCI}$	3.18	18	$R_C C_{JCS}$	0.35
7	$R_{BX} C_{JCX}$	2.98	19	$R_E C_{JCI}$	3.25
8	$R_{BX} C_{JEB}$	0.81	20	$R_E C_{JCX}$	2.89
9	$R_{BX} C_{DE}$	0.96	21	$R_E C_{JEB}$	0.29
10	$R_L C_{JCI}$	0.87 + 0.21F	22	$R_E C_{DE}$	0.37
11	$R_L C_{JCX}$	1.25 + 0.21F	23	$R_E C_{JCS}$	0.17
12	$R_L C_{JEB}$	0.25F	24	$R_E C_L$	0.33

The weighting factors K_i for appropriate RC time constants are summarized in Table 7.1, where F is the fan-out of the circuit. As might be expected, the important time constants such as $R_{BX} C_{JCI}$ have large weighting factors, whereas less critical time constants such as $R_E C_{JCS}$ have much smaller values. A total of 24 terms is needed to fully describe the gate delay. Equation (7.2) is valid provided the parameter values remain approximately constant within the ranges indicated in Table 7.2. Outside these ranges the weighting factors do not remain constant, and the propagation delay expression becomes

Table 7.2. Range of parameter values over which the propagation delay expression is valid (after Chor et al. [3], copyright ©1988 IEEE)

Parameter	Allowed value	
	Minimum	Maximum
R_L	250 Ω	2 kΩ
R_{BI}	0	5 kΩ
R_{BX}	0	1 kΩ
R_C	0	1 kΩ
R_E	0	300 Ω
C_{JCI}	1 fF	0.5 pF
C_{JCX}	1 fF	0.5 pF
C_{JEB}	1 fF	1 pF
C_{JCS}	1 fF	1 pF
C_L	0	1 pF
τ_F	2 ps	100 ps

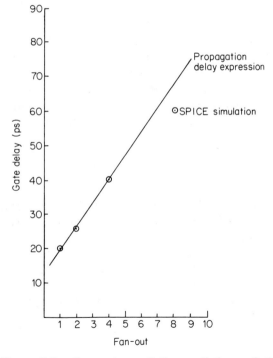

Figure 7.2. Comparison of the predictions of the propagation delay expression with the results of direct SPICE simulations (after Ashburn, et al. [12], copyright ©1987 IEEE)

increasingly inaccurate. Provided this constraint is not violated, equation (7.2) is accurate to approximately $\mp 12\%$ [3].

The gate delay expression is also inaccurate when very high fan-outs are used, or alternatively when the interconnection capacitance C_L is very large. This is illustrated in Figure 7.2, which compares the predictions of equation (7.2) with the results of direct SPICE simulations. It can be seen that the propagation delay expression agrees closely with the SPICE simulations up to a fan-out of about five, but thereafter the agreement is poorer. This discrepancy at high fan-outs is because the propagation delay expression was computed for ring oscillator circuits with unity fan-out.

7.3 CALCULATION OF THE ELECTRICAL PARAMETERS

Having arrived at an analytical expression for the propagation delay in terms of the transistor and circuit parameters the next stage is to relate these parameters to the processing variables such as doping profiles and sheet resistances. In general, this can be achieved through the use of simple device modelling equations, although in some cases more accurate estimates can be obtained by using device-modelling programs such as BIPOLE [5].

The intrinsic and extrinsic base resistances R_{BI} and R_{BX} have already been considered in Chapter 3, and can be calculated using equations (3.58) and (3.57). The dominant component of the emitter resistance is contact resistance, though in polysilicon emitter transistors an additional component is present due to the resistance of the polysilicon/silicon interface [8]. In a well-designed bipolar transistor the collector resistance R_C is determined primarily by the sheet resistance of the buried layer R_{SBL} and also to some extent by the contact resistance R_{CON}. In this case, R_C can be approximated by:

$$R_C = R_{SBL} \cdot \frac{b_c}{l_c} + R_{CON} \qquad (7.3)$$

where b_c and l_c are the width and length of the buried layer. The series resistance of the epitaxial layer may also need to be taken into account in some cases, as illustrated in Figure 3.6.

The load resistance R_L is determined by the operating conditions of the circuit, and in particular by the logic swing ΔV. ECL is a non-saturating logic family, and hence the switching transistors operate in the forward active region. From Figure 7.1 the load resistance is therefore approximately given by:

$$R_L = \frac{\Delta V}{I_C} \qquad (7.4)$$

The collector current is generally chosen so that the transistor is operating close to the peak f_T, since this minimizes the forward transit time.

The zero bias depletion capacitances C_{JCI}, C_{JCX}, C_{JEB} and C_{JCS} can be calculated from the doping profiles using standard textbook *pn* junction theory [9]. The split between intrinsic and extrinsic collector capacitance can be estimated from the relative areas of these two regions, as illustrated in Figure 3.21. Emitter diffusion capacitance C_{DE} is a strong function of the collector current, and can be approximated by [3]:

$$C_{DE} = \frac{2 I_C \tau_F}{\Delta V} \qquad (7.5)$$

The collector diffusion capacitance can be neglected because the switching transistors in Figure 7.1 operate in the forward active region. That is, the collector/base junction is reverse biased. The load capacitance C_L is determined by the capacitance of the interconnections, which can easily be calculated from the field oxide thickness and the area of the metal tracks between the ring oscillator stages.

In theory, the forward transit time τ_F can be calculated from equations (3.39)–(3.41). This gives an approximate value for τ_F, but the use of device-modelling programs is recommended for more accurate estimates. The forward transit time is generally an important component of the gate delay, and hence it is essential that its value is accurately calculated.

7.4 COMPARISON OF CONVENTIONAL AND SELF-ALIGNED PROCESSES

Having computed an expression for the propagation delay in terms of the transistor electrical parameters and related these to the processing variables we are now in a position to investigate the application of the propagation delay expression to process optimization. We will begin by comparing the predicted performance of conventional and self-aligned ECL circuits using the method of analysis described in the previous sections.

Table 7.3. Summary of the process and device properties of a 1 μm silicon bipolar transistor

Process/electrical parameter	Value
Base sheet resistance	3 kΩ/sq
Resistance under the emitter	6 kΩ/sq
Extrinsic base sheet resistance	50 Ω/sq
Buried layer sheet resistance	20 Ω/sq
Contact resistance	2×10^{-7} Ωcm^2
Emitter resistance	1.4×10^{-6} Ωcm^2
Emitter capacitance	2.7×10^{-7} F/cm^2
Collector capacitance	3.2×10^{-8} F/cm^2
Substrate capacitance	1.4×10^{-8} F/cm^2
Peak f_T	6.2 GHz
Forward transit time τ_F	20.8 ps
I_C for peak f_T	1.5×10^4 A/cm^2

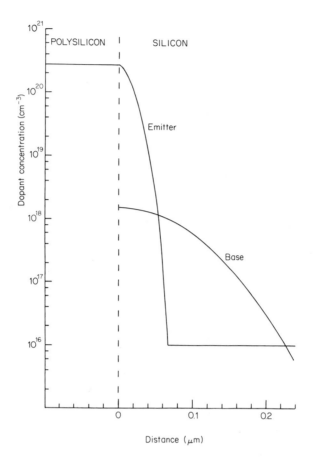

Figure 7.3. Doping profiles for a 6.2 GHz silicon bipolar transistor

The starting point for this comparison is emitter and base doping profiles and transistor layouts for the two types of transistor. In order to facilitate the comparison, a polysilicon emitter has been assumed for both processes, with the emitter and base doping profiles illustrated in Figure 7.3. The profiles are gaussian and the metallurgical junction depths are 0.05 and 0.23 μm for the emitter/base and collector/base junctions, respectively. A base sheet resistance of 3 kΩ/sq and a resistance under the emitter of 6 kΩ/sq are predicted for these profiles, as summarized in Table 7.3. Transistor layouts for the conventional and self-aligned transistor are shown in Figures 7.4 and 7.5. A 1μm process has been assumed, with an alignment tolerance of 0.5 μm and a metal–metal separation of 1.5 μm. The layouts are relatively conservative and, in particular, walled emitters have not been used.

A number of additional process parameters, such as extrinsic base and buried layer sheet resistances, are needed for calculation of the parasitic resistances

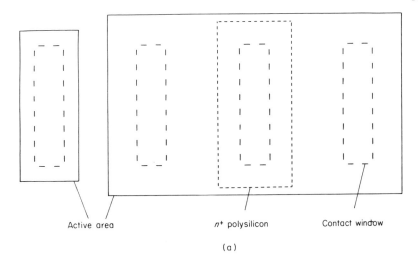

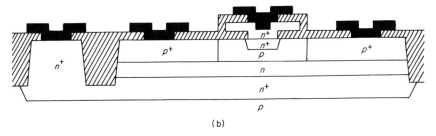

Figure 7.4. Schematic illustration of the layout of an idealized, conventional, silicon bipolar transistor. (a) Plan view; (b) cross-sectional view

of the transistors. Values of 50 Ω/sq and 20 Ω/sq, respectively, have been assumed (Table 7.3), which are typical for silicon bipolar processes of this type. Finally, a contact resistance of $2 \times 10^{-7}\,\Omega cm^2$ [10] and an emitter resistance of $1.4 \times 10^{-6}\,\Omega cm^2$ [8] have been assumed.

The BIPOLE device simulation program [5] has been used to compute the electrical characteristics of the polysilicon emitter bipolar transistor, and the predictions are summarized in the lower half of Table 7.3. For simplicity, a one-dimensional transistor structure has been assumed, so some of the parameters, such as the cut-off frequency, are likely to be slightly optimistic. A cut-off frequency of 6.2 GHz is predicted, which is reasonable for a transistor of this type.

The components of the cut-off frequency are summarized in Table 7.4. The first four terms in the table are the components of the forward transit time τ_F, given in equation (3.39), and the last term T_{RE} is the depletion capacitance term in equation (3.53). The collector resistance term in equation (3.53) has been assumed to be negligible. From the predictions in Table 7.4 it can be seen that all the terms contribute significantly to the total.

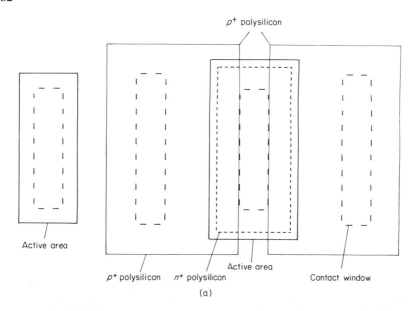

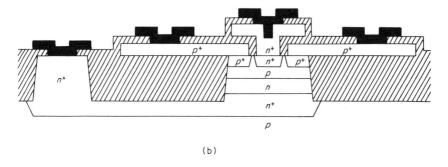

Figure 7.5. Schematic illustration of the layout of an idealized, self-aligned, silicon bipolar transistor. (a) Plan view; (b) cross-sectional view

Table 7.4. Components of the cut-off frequency for the silicon bipolar transistor in Table 7.3

Cut-off frequency f_T (GHz)	Components of f_T (ps)				
	τ_E	τ_{EBD}	τ_B	τ_{CBD}	τ_{RE}
6.2	5.2	1.1	5.2	9.3	4.7

The next stage of the analysis is to use the modelling equations of Section 7.3 to calculate the values of the transistor and circuit parameters required in the propagation delay expression. The results of these calculations are summarized in Table 7.5 for the two types of transistor. As expected, the major advantages of the self-aligned process are considerable reductions in extrinsic

Table 7.5. Transistor electrical parameters for a conventional and a self-aligned 1 μm silicon bipolar transistor (logic swing = 0.4 V)

Parameter	Transistor type	
	Conventional	Self-aligned
R_L	660 Ω	660 Ω
R_{BI}	125 Ω	125 Ω
R_{BX}	257 Ω	13 Ω
R_C	40 Ω	40 Ω
R_E	35 Ω	35 Ω
C_{JCI}	1.3 fF	1.3 fF
C_{JCX}	17.9 fF	4.5 fF
C_{JEB}	10.8 fF	10.8 fF
C_{JCS}	12.5 fF	8.0 fF
C_{DE}	62.4 fF	62.4 fF
C_L	20 fF	20 fF
τ_F	20.8 ps	20.8 ps

Table 7.6. Breakdown of the components of the propagation delay for a conventional and a self-aligned 1 μm silicon bipolar transistor (logic swing = 0.4 V)

Time constant	Delay (ps)	
	Conventional transistor	Self-aligned transistor
τ_F	25.0	25.0
$R_{BI}C_{JCI}$	0.6	0.6
$R_{BI}C_{JCX}$	0.9	0.2
$R_{BI}C_{JEB}$	1.0	1.0
$R_{BI}C_{DE}$	7.6	7.6
$R_{BX}C_{JCI}$	1.1	0.1
$R_{BX}C_{JCX}$	13.7	0.2
$R_{BX}C_{JEB}$	2.2	0.1
$R_{BX}C_{DE}$	15.4	0.8
$R_L C_{JCI}$	0.9	0.9
$R_L C_{JCX}$	17.2	4.3
$R_L C_{JEB}$	1.8	1.8
$R_L C_{JCS}$	1.3	0.8
$R_L C_L$	4.0	4.0
$R_C C_{JCI}$	0.1	0.1
$R_C C_{JCX}$	1.2	0.3
$R_C C_{DE}$	0.8	0.8
$R_C C_{JCS}$	0.2	0.1
$R_E C_{JCI}$	0.1	0.1
$R_E C_{JCX}$	1.8	0.5
$R_E C_{JEB}$	0.1	0.1
$R_E C_{DE}$	0.8	0.8
$R_E C_{JCS}$	0.1	0
$R_E C_L$	0.2	0.2
Total	98.1	50.4

base resistance R_{BX} and extrinsic collector/base capacitance C_{JCX}. The collector/substrate capacitance C_{JCS} is also slightly lower. The reduced extrinsic base resistance is obtained in the self-aligned process because of the small separation between the heavily doped extrinsic base and the polysilicon emitter. The reduced collector/base capacitance is the direct result of the use of the p^+ polysilicon layer to make contact to the base.

The final stage of the analysis is to use the propagation delay expression in equation (7.2) and the weighting factors in Table 7.1 to calculate the components of the gate delay. The results of this procedure are summarized in Table 7.6. These predictions clearly demonstrate the considerable improvement in performance that can be achieved by the use of a self-aligned process. As expected, the lower values of extrinsic base resistance and extrinsic collector/base capacitance are primarily responsible for the improvement. The results in Table 7.6 also clearly demonstrate the way in which the propagation delay expression can be used to identify the dominant time constants. For the self-aligned transistor these can be identified as the forward transit time τ_F and the $R_{BI}C_{DE}$ term.

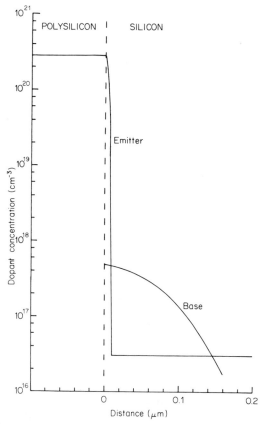

Figure 7.6. Doping profiles for a 17.5 GHz silicon bipolar transistor

7.5 PROCESS OPTIMIZATION

7.5.1 Silicon Bipolar Processes

Having demonstrated the advantages of the self-aligned process we are now in a position to proceed to an investigation of how the process might be further optimized. The list of time constants in Table 7.6 clearly demonstrates that the forward transit time τ_F is by far the dominant component of the gate delay. It is therefore clear that the first stage of the optimization procedure must be to improve the design of the transistor itself.

From the summary of the components of the cut-off frequency in Table 7.4 the collector/base depletion region transit time τ_{CBD} can be identified as the largest term. This can easily be improved by increasing the collector doping concentration, thereby decreasing the depletion width. An added benefit of this change will be a decrease in the transconductance term τ_{RE}. This comes about through the suppression of basewidth-widening effects, which, in turn, leads to a shift of the peak f_T to a higher collector current. Equation (3.53) demonstrates that this will lead directly to an improvement in the cut-off frequency of the transistor. The base transit time is also a significant component of f_T, and can be reduced by decreasing the basewidth or decreasing the base doping (equation (3.40)). The two remaining components of f_T in Table 7.4 are due to the charge stored in the emitter and in the emitter/base depletion region. These can be reduced by decreasing the emitter/base junction depth and decreasing the base doping.

Figure 7.6 shows a more optimum doping profile for the self-aligned transistor in Figure 7.5. The emitter/base junction depth has been reduced to 0.01 μm [11] and the metallurgical basewidth to 0.13 μm. The doping concentration in the base has also been reduced, with the result that the base resistance has increased to 7.3 kΩ/sq, and the resistance under the emitter to 9.7 kΩ/sq, as summarized in Table 7.7. A cut-off frequency of 17.5 GHz is predicted, the components of which are summarized in Table 7.8. A comparison with Table 7.4 shows that all the terms are significantly reduced, but the largest decrease is in the collector/base depletion region transit time τ_{CBD}. It should also be noted that the increase in the cut-off frequency has been achieved at the cost of

Table 7.7. Summary of the process and device properties of an optimized 1 μm silicon bipolar transistor

Process/transistor parameter	Value
Base sheet resistance	7.3 kΩ/sq
Resistance under the emitter	9.7 kΩ/sq
Emitter capacitance	1.8×10^{-7} F/cm^2
Collector capacitance	5.1×10^{-8} F/cm^2
Peak f_T	17.5 GHz
Forward transit time τ_F	5.9 ps
I_C for peak f_T	2.0×10^4 A/cm^2

Table 7.8. Summary of the components of the cut-off frequency for the transistor in Table 7.7

Cut-off frequency f_T (GHz)	Components of f_T (ps)				
	τ_E	τ_{EBD}	τ_B	τ_{CBD}	τ_{RE}
17.5	0.6	0.7	2.9	1.7	3.2

Table 7.9. Transistor parameters for an optimized 1 μm silicon bipolar transistor (logic swing = 0.4 V)

Parameter	Value
R_L	500 Ω
R_{BI}	202 Ω
R_{BX}	13 Ω
R_C	40 Ω
R_E	35 Ω
C_{JCI}	2.0 fF
C_{JCX}	7.1 fF
C_{JEB}	7.2 fF
C_{JCS}	8.0 fF
C_{DE}	23.6 fF
C_L	20 fF
τ_F	5.9 ps

Table 7.10. Breakdown of the components of the propagation delay for an optimized 1 μm silicon bipolar transistor (logic swing = 0.4 V)

Time constant	Delay (ps)
τ_F	7.1
$R_{BI} C_{JCI}$	1.4
$R_{BI} C_{JCX}$	0.6
$R_{BI} C_{JEB}$	1.1
$R_{BI} C_{DE}$	4.6
$R_{BX} C_{JCI}$	0.1
$R_{BX} C_{JCX}$	0.3
$R_{BX} C_{JEB}$	0.1
$R_{BX} C_{DE}$	0.3
$R_L C_{JCI}$	1.1
$R_L C_{JCX}$	5.2
$R_L C_{JEB}$	0.9
$R_L C_{JCS}$	0.6
$R_L C_L$	3.0
$R_C C_{JCI}$	0.2
$R_C C_{JCX}$	0.5
$R_C C_{DE}$	0.3
$R_C C_{JCS}$	0.1
$R_E C_{JCI}$	0.2
$R_E C_{JCX}$	0.7
$R_E C_{JEB}$	0.1
$R_E C_{DE}$	0.3
$R_E C_{JCS}$	0
$R_E C_L$	0.2
Total	29.0

an increased base resistance and collector capacitance. As explained above, the collector current for peak f_T has also increased slightly as a result of the higher collector doping.

Following the procedure described in Section 7.4, the transistor and circuit parameters can be calculated (Table 7.9), along with the components of the propagation delay (Table 7.10). The predicted delay of 29 ps is a considerable improvement on the value in Table 7.6, and demonstrates the effectiveness of the propagation delay expression for process optimization. In the improved self-aligned process in Table 7.10 the dominant components of the delay are the forward transit time τ_F, $R_L C_{JCX}$, $R_{BI} C_{DE}$ and $R_L C_L$. Further improvements in performance could be obtained by reducing the logic swing [12], but this would be at the expense of a degradation in the noise margin.

7.5.2 GaAs/GaAlAs Heterojunction Bipolar Processes

The propagation delay expression is equally applicable to the optimization of GaAs/GaAlAs heterojunction bipolar processes. Using the doping densities in Table 6.2 and a transistor layout similar to that in Figure 6.30, the process and electrical parameters can be computed using the procedure described above. The results are summarized in Table 7.11, along with assumed values for contact resistance [13], emitter resistance [14] and buried layer sheet resistance.

Table 7.11. Summary of the process and device properties of a 1 μm GaAs/GaAlAs heterojunction bipolar transistor

Process/electrical parameter	Value
Base sheet resistance	330 Ω/sq
Resistance under the emitter	330 Ω/sq
Extrinsic base sheet resistance	330 Ω/sq
Buried layer sheet resistance	20 Ω/sq
Contact resistance	1×10^{-6} Ωcm^2
Emitter resistance	2.2×10^{-6} Ωcm^2
Emitter capacitance	1.8×10^{-7} F/cm^2
Collector capacitance	4.5×10^{-8} F/cm^2
Substrate capacitance	0
Peak f_T	45 GHz
Forward transit time	3.0 ps
I_C for peak f_T	2.0×10^4 A/cm^2

Table 7.12. Summary of the components of the cut-off frequency for the GaAs/GaAlAs heterojunction bipolar transistor in Table 7.11

Cut-off frequency f_T (GHz)	Components of f_T (ps)				
	τ_E	τ_{EBD}	τ_B	τ_{CBD}	τ_{RE}
45	0	0	1.2	1.8	0.5

A cut-off frequency of 45 GHz is predicted for the GaAs/GaAlAs heterojunction transistor, which is more than twice the value obtained for the self-aligned silicon transistor. The components of f_T are summarized in Table 7.12, where it can be seen that the improved performance is partly due to the use of a heterojunction emitter and partly to the use of gallium arsenide instead of silicon. For example, the lower values of τ_E and τ_{EBD} are obtained as a result of the suppression of hole injection into the emitter by the heterojunction. In contrast, the low value of base transit time τ_B is a direct result of the high electron mobility of gallium arsenide (Figure 2.6).

A comparison of the process parameters in Tables 7.7 and 7.11 reveals further advantages of GaAs/GaAlAs technology. First, the improved gains of heterojunction transistors can be traded for very low values of resistance under the emitter. This leads directly to an extremely small intrinsic base resistance R_{BI}, as illustrated in Table 7.13. The lower intrinsic base resistance is, to some extent, offset by a higher extrinsic base sheet resistance, as shown in Table 7.11. Fortunately, however, this disadvantage is not too serious in a self-aligned heterojunction process because the base contact is immediately adjacent to the emitter (Figure 6.30(d)). The final value of extrinsic base resistance R_{BX} in Table 7.13 is therefore only slightly higher than the value obtained for silicon in Table 7.9. The emitter and collector depletion capacitances of the heterojunction process are very similar to those obtained for silicon, as can be seen by comparing Tables 7.7 and 7.11. However, it should be noted that the emitter capacitance of $1.8 \times 10^{-7}\,F/cm^2$ for the heterojunction process is obtained with a much higher base doping than in the silicon process. This has been achieved by trading the improved gain of the heterojunction transistor for a lower emitter doping.

The components of the propagation delay for the GaAs/GaAlAs heterojunction process are summarized in Tale 7.14, A delay of 19.3 ps is predicted

Table 7.13. Transistor parameters for a 1 μm GaAs/GaAlAs heterojunction bipolar transistor (logic swing = 0.4 V)

Parameter	Value
R_L	500 Ω
R_{BI}	7 Ω
R_{BX}	33 Ω
R_C	40 Ω
R_E	55 Ω
C_{JCI}	1.8 fF
C_{JCX}	8.1 fF
C_{JEB}	7.2 fF
C_{JCS}	0
C_{DE}	12.0 fF
C_L	20 fF
τ_F	3.0 ps

Table 7.14. Breakdown of the components of the propagation delay for a 1 μm GaAs/GaAlAs heterojunction bipolar transistor (logic swing = 0.4 V)

Time constant	Value
τ_F	3.6
$R_{BI}C_{JCI}$	0
$R_{BI}C_{JCX}$	0
$R_{BI}C_{JEB}$	0
$R_{BI}C_{DE}$	0.1
$R_{BX}C_{JCI}$	0.2
$R_{BX}C_{JCX}$	0.8
$R_{BX}C_{JEB}$	0.2
$R_{BX}C_{DE}$	0.4
$R_L C_{JCI}$	1.0
$R_L C_{JCX}$	5.9
$R_L C_{JEB}$	0.9
$R_L C_{JCS}$	0
$R_L C_L$	3.0
$R_C C_{JCI}$	0.2
$R_C C_{JCX}$	0.5
$R_C C_{DE}$	0.2
$R_C C_{JCS}$	0
$R_E C_{JCI}$	0.3
$R_E C_{JCX}$	1.3
$R_E C_{JEB}$	0.1
$R_E C_{DE}$	0.2
$R_E C_{JCS}$	0
$R_E C_L$	0.4
Total	19.3

compared with 29 ps for silicon. The dominant components of the gate delay are $R_L C_{JCX}$, $R_L C_L$ and τ_F. These results indicate that further improvements of performance could be obtained by reducing the extrinsic collector capacitance, the forward transit time and the logic swing [12]. The use of proton implantation to produce a buried isolation region [15] provides an effective means of decreasing the collector/base capacitance of GaAs/GaAlAs heterojunction transistors. There is also some scope for decreasing the basewidth and increase the collector doping of the transistor, thereby reducing the forward transit time. In this context, cut-off frequencies as high as 75 GHz have been reported in the literature [16]. Theoretical comparisons of the performance of silicon and GaAs/GaAlAs heterojunction processes [12,17] indicate that the performance advantage of the latter technology is between a factor of 1.5 and 3.

References

1. G. W. Taylor and J. G. Simmons, 'Figure merit for integrated bipolar transistors', *Solid State Electronics*, **29**, 941 (1986).

2. A. D. Welbourn, 'Gigabit logic', *IEE Proc.*, Pt I, **129**, 157 (1983).
3. E. F. Chor, A. Brunnschweiler and P. Ashburn 'A propagation delay expression and its application to the optimization of polysilicon emitter ECL processes', *IEEE Jnl Solid State Circuits*, **SC23**, No. 1 (1988).
4. D. A. Antoniadis, S. E. Hansen, R. W. Dutton and A. B. Gonzalez, 'SUPREM I: a program for IC process modeling and simulation', *Tech. Rept 5019-1*, Stanford Electronics Laboratory, Stanford University (1977)
5. D. J. Roulston, S. G. Chamberlain and J. Sehgal, 'Simplified computer aided analysis of double diffused transistors including two dimensional high-level effects', *IEEE Trans. Electron. Devices*, **ED19**, 809 (1972).
6. D. D. Tang and P. M. Solomon, 'Bipolar transistor design for optimized power-delay logic circuits', *IEEE Jnl Solid State Circuits*, **SC14**, 679 (1979).
7. R. Ranfft and H. M. Rein, 'A simple optimization procedure for bipolar subnanosecond ICs with low power dissipation', *Microelectronics Jnl*, **13**, 23 (1982).
8. E. F. Chor, P. Ashburn and A. Brunnschweiler, 'Emitter resistance of arsenic and phosphorus doped polysilicon emitter transistors', *IEEE Electron. Device Lett.*, **EDL6**, 516 (1985).
9. S. M. Sze, *Physics of Semiconductor Devices*, John Wiley, Chichester (1981), p. 74.
10. R. L. Maddox, 'Optimization of VLSI contacts', *IEEE Trans. Electron. Devices*, **ED32**, 682 (1985).
11. A. Cuthbertson and P. Ashburn, 'An investigation of the trade-off between enhanced gain and base doping in polysilicon emitter bipolar transistors', *IEEE Trans. Electron. Devices*, **ED32**, 2399 (1985).
12. P. Ashburn, A. A. Rezazadeh, E. F. Chor and A. Brunnschweiler, 'Comparison of silicon bipolar and GaAs/GaAlAs heterojunction technologies for high-speed ECL circuits', *IEEE Bipolar Circuits and Technology Meeting*, 61 (1987).
13. S. Tiwari, 'GaAs/GaAlAs heterojunction bipolar transistors: prospects, progress and problems', *IEEE Bipolar Circuits and Technology Meeting*, 21 (1986).
14. M. F. Chang, P. M. Asbeck, D. L. Miller and K. C. Wang, 'GaAs/GaAlAs heterojunction bipolar transistors using a self-aligned substitutional emitter process', *IEEE Electron. Device Lett.*, **EDL7**, 8 (1986).
15. O. Nakajima, K. Nagata, Y. Yamauchi, H. Ito and T. Ishibashi, 'High-speed AlGaAs/GaAs HBTs with proton implanted buried layers', *IEDM Technical Digest*, 266 (1986).
16. K. C. Wang, P. M. Asbeck, M. F. Chang, G. J. Sullivan and D. L. Miller, 'High speed circuits for lightwave communication systems implemented with AlGaAs/GaAs heterojunction bipolar transistors', *Bipolar Circuits and Technology Meeting Digest*, 142 (1987).
17. M. Kurata, R. Katoh, J. Yoshida and J. Akagi, 'A model base comparison: GaAs/GaAlAs HBT versus silicon bipolar', *IEEE Trans. Electron. Devices*, **ED33**, 1413 (1986).

APPENDIXES

APPENDIX I BIPOLAR TRANSISTOR MODEL PARAMETERS

DC Ebers–Moll Model

β_F	Forward common emitter current gain
β_R	Reverse common emitter current gain
I_S	Saturation current

AC Ebers–Moll Model

C_{JEB}	Emitter/base junction capacitance
C_{JBC}	Base/collector junction capacitance
C_{JCS}	Collector/substrate junction capacitance
R_B	Base resistance
R_C	Collector resistance
R_E	Emitter resistance
τ_F	Forward transit time
τ_R	Reverse transit time

Gummel–Poon Model

B	Parameter for modelling base widening effects
I_{KF}	Forward knee current, defining the onset of high-level injection
I_{KR}	Reverse knee current, defining the onset of high-level injection
I_{SE}	Foward saturation current for recombination in the depletion region
I_{SC}	Reverse saturation current for recombination in the depletion region
N_E	Forward, low-current ideality factor
N_C	Reverse, low-current ideality factor
V_{AF}	Forward Early voltage
V_{AR}	Reverse Early voltage

Small-Signal Hybrid-π Model

C_μ	Collector/base capacitance
C_π	Emitter/base capacitance
g_m	Transconductance
r_π	Input resistance representing the linearized emitter/base diode

APPENDIX 2 FUNDAMENTAL PHYSICAL CONSTANTS

Quantity	Value
Boltzmann's constant (K)	1.38×10^{-23} J/°K
Electronic charge (q)	1.602×10^{-19} C
Permittivity of free space (ε_o)	8.85×10^{-12} C^2/N – m
Planck's constant (h)	6.626×10^{-34} J – s
Free electron mass (m_o)	9.108×10^{-31} kg
1 electron volt (eV)	1.602×10^{-19} J
Thermal voltage (300 K) (KT/q)	0.0258 V
Angstrom unit (Å)	1×10^{-10} m

APPENDIX 3 PROPERTIES OF SILICON AND GALLIUM ARSENIDE

Quantity	Silicon	Gallium arsenide
Lattice constant (300 K) (Å)	5.431	5.653
Atoms/cm^2	5×10^{22}	4.42×10^{22}
Melting point (°C)	1415	1238
Bandgap (eV) (0 K)	1.170	1.519
Bandgap (eV) (300 K)	1.125	1.424
Electron mobility (cm^2/V – s)	1500	8500
Hole mobility (cm^2/V – s)	450	400
Intrinsic carrier concentration n_{io} (300 K) (cm^{-3})	1.45×10^{10}	1.79×10^6
Effective density of states in the conduction band N_c (cm^{-3}) (300 K)	2.8×10^{19}	4.7×10^{17}
Effective density of states in the valence band N_v (cm^{-3}) (300 K)	1.04×10^{19}	7.0×10^{18}
Dielectric constant ε_r	11.9	13.1
Electron affinity χ (V)	4.05	4.07

APPENDIX 4 PROPERTIES OF SILICON DIOXIDE

Quantity	Value
Molecules/cm^3	2.3×10^{22}
Melting point (°C)	$\simeq 1600$
Dielectric constant ε_r	3.9
Bandgap (eV)	$\simeq 9$
Breakdown field (V/μm)	600

INDEX

abrupt heterojunction 125
access time 159
AC Ebers–Moll model 63
A/D converter 166
activate 116, 144, 148
active transistor area 80
alignment 135
alignment tolerance 180
alloy 155
alloy spikes 153
aluminium 155
aluminium spikes 153
amorphization 145
amorphous silicon 89, 115
amplifier 11
analogue bipolar process 1, 153
analogue circuit 67, 134, 153, 166
anisotropic etching 159
anneal 144, 148
antimony 135, 153
apparent bandgap narrowing 31, 38, 129
arsenic
  buried layer 135, 153
  emitter 148
  polysilicon emitter 89, 115, 116
arsenic segregation 95, 116
Auger recombination 35
autodoping 135
avalanche breakdown 53

balling 111
ballistic transport 130
band-bending 118
band discontinuities 127
bandgap engineering 130
bandgap narrowing 31, 38, 129
base bandgap grading 130
base current
  components 14
  hole diffusion current 20
  in a heterojunction emitter 125, 126, 128
  in a polysilicon emitter 96, 107
  in a shallow emitter 25, 40, 91, 148
  incorporation of heavy doping effects 38
  recombination current in the base 23
  recombination current in the depletion region 46
base fabrication 144
base Gummel number 25, 39
base resistance 63, 80, 159, 178, 183, 188
base storage time 66, 74, 79, 181, 185, 188
base transit time 66, 74, 79, 181, 185, 188
base transport factor 16
base-widening 71, 75, 79, 86
basewidth modulation 24, 50, 69
BF_2 145
BICMOS 3, 166
bird's beak 140, 168
boron 144, 159
boundary condition 20, 22, 26
breakdown 10, 51
  avalanche 53
  punch-through 52
  soft 145, 150
  Zener 52
breakdown voltage 10, 51
  common base BV_{CBO} 53
  common emitter BV_{CEO} 54
built-in electric field 74, 131
built-in voltage 66, 126
buried layer 1, 135, 167, 178

capacitance/voltage relationship 66
capacitor fabrication 157, 168
cap layer 165
capture cross-section 44, 45, 100, 104

carrier concentration 6
channelling 145
channel-stop 140, 142
charge storage 63, 66
chemical vapour deposition 145, 159
CMOS 139, 166
collector/base capacitance 65, 82, 159, 165, 178, 183, 188
collector/base depletion region transit time 66, 74, 79, 181, 185, 188
collector current 24, 39
collector diffusion capacitance 179
collector diffusion region 78, 135
collector resistance 49, 64, 78, 135, 167, 178, 183, 188
collector/substrate capacitance 65, 163
columnar grain structure 115
common base connection 9, 51, 54
common base current gain 8, 16
common collector connection 9
common emitter connection 9, 51, 54
common emitter current gain 7, 24, 27, 33, 42, 48, 125, 128
compensation 155
concentration-dependent oxidation 135
conduction band discontinuity 127, 130
conductivity modulation 48
contact
　alloy 155
　resistance 64, 81, 178, 187
　window 155
continuity equation 16, 19
critical electric field 53
crystallographic defect 136, 140, 143, 145, 147, 149
current crowding 82, 86
current gain 7, 24, 27, 33, 42, 48, 125, 128
current mirror circuit 156
current mode logic 159, 166
cut-off 6
cut-off frequency f_T 76, 156, 162, 181, 185
CVD 142, 145, 159

dangling bond 95, 97, 99
DC Ebers–Moll model 59
deep level 44
defect generation 135, 143, 144
defect nucleation 136
degenerate 40, 107
densification 145
depletion capacitance 63, 65
depletion width 126
deposited oxide 142, 145, 159

deposition 144, 159
device optimization 174, 179, 185
device simulation 175, 178, 181
dichlorosilane 136
dielectric constant 17, 126
diffusion of dopant 1
diffusion capacitance 63, 66, 68, 179, 183, 188
diffusion coefficient 17, 116, 135
diffusion current 17, 19, 20, 25
diffusion length
　of electrons 22, 39
　of holes 21, 37
diffusivity 17
digital bipolar process 159, 163
digital circuits 2, 67, 134, 166, 174
diode 157
dislocation
　base 145
　emitter 147, 149
　epitaxy 136
　isolation 140, 143
dislocation network 147
dopant segregation 95, 116
doping profile 117, 145, 180, 185
double heterojunction 131
drift current 17, 75
drift velocity 75
drive-in 145, 147
dry oxygen 135

Early voltage 51, 71
Ebers–Moll model 59, 63
ECL 3, 159, 175
edge dislocations 147
effective barrier height 95, 102, 107
effective doping concentration 32
effective recombination velocity 96
Einstein relation 17
electrically active dopant 116
electric displacement 126
electron
　capture 44
　capture cross-section 44
　concentration 6
　emission 45
electron affinity 125, 126
emission probability 45
emitter/base capacitance 65, 92, 124, 179, 183, 188
emitter/base depletion region delay 66, 74, 79, 181, 185, 188
emitter/collector pipes 140, 148
emitter current crowding 82

emitter delay 66, 74, 79, 181, 185, 188
emitter deposition 146
emitter diffusion capacitance 66, 179, 183, 188
emitter dip effect 148, 151
emitter drive-in 89, 110, 145, 147
emitter edge dislocations 147
emitter efficiency 16
emitter fabrication 89, 146
emitter Gummel number 40
emitter implant 89, 110, 147
emitter push effect 148, 151
emitter resistance 50, 64, 107, 178, 183, 187
epitaxial layer 1, 78, 135, 167
epitaxial regrowth 111
epitaxial stacking faults 136
epitaxy 1, 135
etch-stop 142, 162
evaporation 137
extrinsic base resistance 80, 159, 165, 178, 183, 186, 188
extrinsic collector/base capacitance 83, 159, 165, 179, 183, 186, 188

facet 139
fan-out 131, 174, 176
Fermi level 5, 20
field oxide 179
figure of merit 80, 174
fillet 159
f_{MAX} 80, 174
forward active region of operation 6, 60, 62
forward Early voltage 71
forward knee current 71
forward transit time 71
forward transit time 67
   components of 74, 79, 181, 185, 188
   in the gate delay expression 175, 179, 183, 188
   variation with I_C 71, 75
   relationship to f_T 76
frequency divider 159, 166
f_T 76, 156, 162, 166, 168, 174, 181, 185, 188

GaAs/GaAlAs heterojunction bipolar process 4, 163
GaAs/GaAlAs heterojunction bipolar transistor 123, 126, 187
GaInAsP 123
gain 7, 24, 27, 33, 42, 48, 125, 128
gain degradation 145, 149, 162

gas-phase doping 137
gate delay 162, 166, 174, 182, 187, 188
gate delay expression 175, 182, 187, 188
generation current 54, 149
generation rate 17, 45
generation/recombination centre 44
generation/recombination in the emitter/base depletion region 42, 46, 73, 128, 149
graded base 130
graded heterojunction 130
grain boundary 89, 97, 114, 116
grain boundary blocking mechanism 95, 97
grain boundary diffusion 116
grain growth 115
groove isolation 141
Gummel number 25, 39, 40
Gummel plot 10, 23, 43, 49, 65, 70, 93, 112, 119
Gummel–Poon model 69

Hall profiling 116
heavy doping effects 27, 38
heterojunction 123
   abrupt 125, 126
   graded 130
heterojunction bipolar process 3, 163
heterojunction bipolar transistor 3, 123, 125, 126, 130, 187
heterojunction collector 131
heterojunction emitter 123, 125, 126, 130, 187
HF surface treatment 83, 111
high-current gain 48
high-level injection 48, 69
high resolution transmission electron microscopy 110
hole
   capture 45
   capture cross-section 45, 100, 103
   concentration 6
   emission 45
hole diffusion current 20, 25, 40, 98, 106, 125
hybrid-Π model 61
hydrofluoric acid 93, 111
hydrogen alloy 155

ideality factor 42, 47, 73, 84
impact ionization 53
implantation
   base 144
   buried layer 135

implantation (*cont.*)
  emitter 146
  polysilicon 89, 110, 114
implantation anneal 144, 148
implantation damage 144, 147
InP 123
input resistance r_Π 68
integrated injection logic I^2L 3, 131, 163
interconnection capacitance 176, 179
interfacial layer 93
  break-up 111
  electrical properties 102, 107
  physical structure 110
intrinsic base resistance 80, 92, 123, 178, 183, 186, 188
intrinsic base sheet resistance 81, 179, 185, 187
intrinsic carrier concentration 6, 21, 32, 34, 125
intrinsic collector/base capacitance 83, 179, 183, 186, 188
intrinsic Fermi level 5, 20, 45
inverse active region of operation 6, 60, 62
inverse gain 131
inversion layer 140, 142
ion implanted resistors 159
isolation 2, 139, 153, 162, 167

JFET 159
junction breakdown 10, 51, 145, 150
junction isolation 1, 139, 153

kink region 119
knee current 71

large-signal transistor model 59
latch-up 167
lateral oxidation 140
lateral *pnp* transistor 157, 163
lattice mismatch 123, 128
laser trimming 159
layout 180
leakage current 53, 145, 149
lifetime 19, 35, 36, 38
lift-off 165
linearity of resistors 157
liquid phase epitaxy 4
load capacitance 166, 174, 179, 183, 188
load resistance 178, 183, 188
local oxidation 139
logic swing 178, 183, 186, 187, 188
looping dislocations 147
low-current gain 42, 46, 73, 128, 149, 156
low-level injection 19

low pressure chemical vapour deposition LPCVD 142, 149, 159

majority carrier charge in the base 69
majority carrier mobility 27
matching 156
merged transistor logic 3, 131, 163
metal film resistor 159, 168
metal organic chemical vapour deposition MOCVD 4, 123, 165
metal precipitates 150
metallic contamination 149
metallization problems 140, 153
Miller effect 82
minority carrier mobility 30
MIS device 117
mobility 17, 27
model parameters 58
  AC Ebers–Moll model 63, 66, 67
  DC Ebers–Moll model 59
  gate delay expression 176, 178
  Gummel–Poon model 72
  low-current gain modelling 73
  non-linear hybrid-Π model 61
  SPICE model 84
modelling bipolar transistors 58
modified Richardson constant 107
molecular beam epitaxy MBE 4, 123, 165
molecular boron BF_2 145
mole fraction 128
MOS capacitor 157
MOSFET 166
multiplication factor 54

n^+ collector diffusion 78, 135
negative resistance kink 119
noise 159, 166
noise margin 166, 187
non-ideal diode 73
non-linear hybrid-Π model 61
non-saturating logic 178
non-threshold logic 159
n-well CMOS 166

ohmic contact 26, 155, 165
one-dimensional bipolar transistor 17
opaque emitter 20, 42
optimization 174, 179, 185
optoelectronic integration 4, 131
out-diffusion 137
output conductance 51
oxidation rate 135
oxidation drive-in 135, 145, 147
oxide balling 111

oxide isolation 2, 139, 159, 167
oxide spacer 159
oxygen implantation 165

packing density 138, 140, 163
pad oxide 140
parameters for transistor models 58
  AC Ebers–Moll model 63, 66, 67
  DC Ebers–Moll model 59
  gate delay expression 176, 178
  Gummel–Poon model 72
  low-current gain modelling 73
  non-linear hybrid-Π model 61
  SPICE model 84
parasitic capacitance 134, 140, 176
parasitic resistance 176
parasitic transistor 158, 167
partial pressure 145
pattern shift 138
peripheral capacitance 27, 92, 162
permittivity 17, 126
phosphorus 115, 146, 151, 155, 167
phosphorus-vacancy pair 151
pinch resistor 81, 156
pipe 140, 148
planar process 1
plasma etch 140
platinum silicide 159
p^+n^+ junction 162
pn product 6, 21, 32, 34
pnp transistor 157, 163, 167
Poisson's equation 17, 75, 126
polycrystalline silicon 89
  deposition 115
  modelling of conduction 97
  physical structure 114
polysilicon-contacted emitter 91
polysilicon emitter 27, 89
  basic physics 92
  emitter resistance 107
  fabrication 90, 159
  practice 109
  SIS emitter 117
  theory 96
polysilicon resistor 168
polysilicon/silicon interface 92
potential barrier 103, 107
power device 23, 42, 49, 137, 168
predeposition 144, 159
process modelling 175
process optimization 174, 179, 185
profile 117, 145, 180, 185
propagation delay 162, 166, 174, 182, 187, 188

propagation delay expression 175, 182, 187, 188
proton implantation 165, 189
punch-through 52

quasi-equilibrium 20, 46
quasi-Fermi level 20, 41, 102
quiescent operating point 67

RAM 159
rapid thermal annealing 165
RCA clean 111
reactive ion etching 142, 159
recessed silicon dioxide 139, 162
recombination 17, 19
  Auger 35
  via deep levels 44
recombination centre 44, 46, 149
recombination current
  in the base 14, 21
  in the emitter 95, 106
  in the emitter/base depletion region 46, 73
  at grain boundaries 99
  at the polysilicon/silicon interface 95, 103
recombination in the base 14, 21
recombination in the emitter/base depletion region 14, 42, 44, 47, 73, 128, 149
recombination rate 17, 19, 45, 46
recombination velocity 26, 96
  at grain boundaries 99
  at interface states 103
  at the polysilicon/silicon interface 96, 103
  at the polysilicon surface 101
  at the surface 26, 98
recombination via deep levels 44
regrowth 111
relative permittivity 17
resistance under the emitter 81, 179, 185, 187
resistor 153, 156, 168
reverse active region of operation 7, 60, 62
reverse current 149
reverse Early voltage 71
reverse I/V characteristic 149
reverse knee current 71
reverse transit time 67, 175
Richardson constant 107
ring oscillator 166, 174, 178

saturation 6, 61
saturation current 9, 59, 61, 83
saturation voltage 135
scaling 52, 92
scattering limited velocity 75

scattering mechanisms 27
Schottky diode 155, 159
seeded growth 136
segregation 95, 116
selective epitaxy 139, 143
selective etch 162, 165
self-aligned 2, 4, 159, 163
self-aligned bipolar process
  GaAs/GaAlAs 163, 187
  silicon 159, 180, 185
semi-insulator 163
series resistance 50, 63
shallow base 145
shallow emitter 25, 40, 91, 148
sheet resistance 29
  base 80, 145, 153
  buried layer 135, 178
  emitter 155
  n-well 167
Shockley, Read, Hall recombination 44
sidewall capacitance 27, 92, 140, 162
silane 136
silicon carbide 4, 123
silicon implantation 145
silicon nitride 140, 165
silicon tetrachloride 136
SIMS 116
single-crystal emitter 106
SIPOS 4, 123
SIS emitter 117
small-signal current gain 11
small-signal hybrid-Π model 67, 77
small-signal operation 67, 77
soft breakdown 150
solid solubility 145
spacer 159
SPICE model 58, 83, 177
spiking 153
stacking faults 136
stress relief oxide 140
substrate capacitance 65, 163, 179, 183, 188
super-β transistor 159
surface concentration 135, 145
surface recombination 26, 155
surface treatment 93, 110

τ_F 67
  components of 74, 79, 181, 185, 188
  gate delay expression 175, 179, 183, 185, 188
  relationship to f_T 76
  variation with I_C 71, 75
temperature coefficient of resistance 156

temperature dependence
  of the current gain 129
  of the leakage current 150
thermal oxidation 135, 139, 145, 147, 153
thermal velocity 44, 100, 103
thermionic emission 103
thin film resistor 159, 168
threshold voltage 167
time constant 175
transconductance 11, 68, 166
transistor layout 180
transistor optimization 174
transistor parameters 58
  AC Ebers–Moll model 63, 66, 67
  DC Ebers–Moll model 59
  gate delay expression 176, 178
  Gummel–Poon model 72
  low-current gain modelling 73
  non-linear hybrid-Π model 61
  SPICE model 84
transistor simulation 175, 178, 181
transmission electron microscopy 110, 114
transparent emitter 25, 40, 106
transport factor 16
traps 44, 100
trench isolation 141
triple diffused bipolar transistor 139, 166
tunnelling breakdown 52, 162
tunnelling mechanism in polysilicon emitters 94, 103, 107
tunnelling via traps 118
two-dimensional modelling 82

unity current gain frequency f_T 76, 156, 162, 174, 181, 185, 188
unity power gain frequency f_{MAX} 80, 174

vacuum level 124, 127
valence band discontinuity 125, 127
velocity of carriers
  drift 75
  thermal 44, 100, 103
  scattering limited 75

walled emitter 140, 180
washout 138
weighting factor 175
well fabrication 166
wet oxidation 145, 147, 155
wide-bandgap emitter 123, 125

yield problems in bipolar processes 148, 165

Zener breakdown 51
Zener diode 157